AF614771

Beta Cells in Health and Disease

Edited by Shahzad Irfan and Haseeb Anwar

Published in London, United Kingdom

Beta Cells in Health and Disease
http://dx.doi.org/10.5772/intechopen.1001550
Edited by Shahzad Irfan and Haseeb Anwar

Contributors
Michael Awuku, Sheila Owens-Collins, Jessie M. Barra, Holger A. Russ, Susan Giblin, Clodagh O'Gorman, Shahzad Irfan, Humaira Muzaffar, Imran Mukhtar, Farhat Jabeen, Haseeb Anwar

First published in London, United Kingdom, 2024 by IntechOpen
IntechOpen is the global imprint of INTECHOPEN LIMITED, registered in England and Wales, registration number: 11086078, 5 Princes Gate Court, London, SW7 2QJ, United Kingdom

British Library Cataloguing-in-Publication Data
A catalogue record for this book is available from the British Library

Additional hard and PDF copies can be obtained from orders@intechopen.com

Beta Cells in Health and Disease
Edited by Shahzad Irfan and Haseeb Anwar
p. cm.
Print ISBN 978-1-83769-628-4
Online ISBN 978-1-83769-627-7
eBook (PDF) ISBN 978-1-83769-629-1

For EU product safety concerns:
IN TECH d.o.o., Prolaz Marije Krucifikse Kozulić 3, 51000 Rijeka, Croatia,
info@intechopen.com or visit our website at intechopen.com.

Meet the editors

Shahzad Irfan is an assistant professor and researcher in the Department of Physiology, Govt. College University Faisalabad, Pakistan. He obtained his Ph.D. in Endocrinology and Physiology of Reproduction in 2014 as a DAAD fellow from the University of Münster, Germany, after which he moved to COMSATS University, Islamabad, Pakistan as an assistant professor in the Department of Biosciences. Since joining the Department of Physiology in 2018, Dr. Irfan has been actively involved in the pathobiology of diabetes mellitus with functional studies on in vitro models of pancreatic beta cells and in vivo models of type 1 and type 2 diabetes. His focus has been on the specific impact of different known herbal extracts and phytochemicals on the regenerative capacity of pancreatic beta cells, insulin synthesis, and secretion, and transcriptional modulation of hepatic glycolytic, lipolytic, gluconeogenic, and lipogenic genes through different transcription factors in type 1 and type 2 rodent and primate models.

Dr. Haseeb Anwar is an associate professor and chair of the Department of Physiology, Govt. College University Faisalabad, Pakistan. He is also a founding member of the university's Department of Physiology. He obtained his Ph.D. from the Department of Physiology, University of Agriculture Faisalabad, Pakistan, in 2012. His research focuses on the pathobiology of different metabolic diseases and microbial gut physiology. He has developed and validated different in-house experimental protocols ranging from in vitro enzyme activity modules to the measurement of total antioxidant capacity (TAC) and total oxidant status (TOS) in different sample types. Dr. Anwar has won numerous competitive research grants from funding bodies. He has also received multiple excellence awards, including Best Researcher awards from different scientific organizations. He has published more than 100 peer-reviewed research articles in international scientific journals in the last 10 years.

Contents

Preface **IX**

Chapter 1 **1**
Understanding Insulin: A Primer
by Michael Awuku

Chapter 2 **13**
Type 1 Diabetes and Beta Cells
by Sheila Owens-Collins

Chapter 3 **33**
Stem Cell-Derived Pancreatic Beta Cells for the Study and Treatment of Diabetes
by Jessie M. Barra and Holger A. Russ

Chapter 4 **51**
Physical Activity in Children and Adolescents with Type 1 Diabetes
by Susan Giblin and Clodagh O'Gorman

Chapter 5 **63**
Impaired Physiological Regulation of ß Cells: Recent Findings from Type 2 Diabetic Patients
by Shahzad Irfan, Humaira Muzaffar, Imran Mukhtar, Farhat Jabeen and Haseeb Anwar

Preface

The pancreatic beta cells present in the islets of Langerhans are the only source of insulin in the body. Insulin is a key metabolic hormone and plays a central role in the maintenance of glucose homeostasis, which includes the regulation of glucose uptake in muscle and adipose tissue as well as carbohydrate, fat, and protein metabolism in the liver. Defective beta cell functioning results in subnormal plasma levels of insulin, leading to either hyperglycemia (diabetes mellitus) or hypoglycemia. Hypoglycemia is a rare, although life-threatening condition, in which the beta cell's hyperactivity results in unnecessarily high insulin levels. Contrary to hypoglycemia, diabetes mellitus is a very common disorder turning into an epidemic. Diabetes is profoundly considered a major threat to human health in the current century. The number of diabetic patients worldwide is rapidly increasing and is predicted to reach around 650 million by 2040, according to the International Diabetes Federation (IDF). Diabetes mellitus is a heterogeneous disorder with some forms such as maturity-onset diabetes of the young and permanent neonatal diabetes being primary genetic disorders of the beta cell. The common forms of diabetes mellitus, designated type 1 and type 2, are multifactorial in origin with both genetic and environmental factors contributing to their development. Type 1 diabetes is caused by the autoimmune destruction of beta cells leading to an absolute deficiency of insulin and fatal hyperglycemia and ketoacidosis if not treated. Type 2 diabetes is a disorder of relative deficiency of insulin resulting when the beta cell mass is not able to expand and thereby secrete more insulin in response to an increasing demand such as occurs in obese individuals. In every form of diabetes, the pancreatic beta cell plays a central role in the disease process. Understanding beta cell biology and regulation is fundamental for the development of treatment options for diabetics and might help us to formulate diabetes preventive measures in everyday life. This book provides a comparative, up-to-date review of beta cell biology in human health and disease. The chapters address the architecture and pathology of normal and diabetic pancreatic islets; regulation of beta cell proliferation and death; the potential of stem cells as beta cell replacement therapies; insulin biosynthesis from transcription to processing; and regulation of insulin secretion, including metabolic control of the beta cell. We hope this book inspires students and young investigators to understand and get involved in diabetes research and develop new approaches for the prevention and treatment of diabetes. In addition, we hope it will provide encouragement to others in the field. Although we have made a great deal of progress in understanding the relationship between the beta cell and health and disease, much remains to be done. We gratefully acknowledge the contributing authors for their chapters.

Shahzad Irfan, Ph.D.
Faculty of Life Sciences,
Department of Physiology,
Government College University Faisalabad,
Faisalabad, Pakistan

Dr. Haseeb Anwar
Associate Professor,
Department of Physiology,
Government College University Faisalabad,
Faisalabad, Pakistan

Chapter 1

Understanding Insulin: A Primer

Michael Awuku

Abstract

Insulin is an essential protein hormone secreted by the beta cells of the islet of Langerhans in the pancreas which is involved in glucose homeostasis, cell metabolism and mitogenesis. It is essential that healthcare providers are conversant with the normal physiology of this protein in the human body, to aid them in clinical decision-making when processes pertaining to this crucial substance go aberrant in the "corpus". Healthcare providers will then be able to better appreciate the pathophysiology of disease states pertaining to this hormone most importantly diabetes which is a great cause of global morbidity and mortality. Healthcare providers should be comfortable especially in recognizing these disease states clinically and instituting the most appropriate form of management in-line with the most recent evidence-based medicine to improve patient outcomes.

Keywords: insulin, diabetes, beta cells, pancreas, hyperglycemia, hypoglycemia

1. Introduction

Insulin (from the Latin word *insula*, meaning island) is an essential protein hormone secreted by the beta cells of the islet of Langerhans in the pancreas which is involved in glucose utilization, cell metabolism and mitogenesis [1, 2]. It is essential that healthcare providers are conversant with the normal physiology of this protein in the human body, to aid them in clinical decision-making when processes pertaining to this crucial substance go aberrant in the "corpus".

Diabetes mellitus (DM) needs no introduction as a significant cause of morbidity and mortality. In 2015, it was estimated that 415 million people were afflicted by diabetes, this number is expected to rise to 642 million in 2040, with an increasing disease burden being shared by people in the developing world. The microvascular and macrovascular complications of this disease have plagued man for a long time. These negative consequences include nephropathy, neuropathy, retinopathy, stroke, renovascular disease, limb ischemia and the dreaded amputation of the limb [3, 4]. The earliest references to this disease in written literature go as far back as 1550 BCE [5]. Despite great progress in our understanding of diabetes and our ever growing arsenal of therapeutic agents used to combat the disease, this old foe of man remains very much alive.

Currently, diabetic kidney disease (DKD) is the leading cause of end-stage renal failure in the world. Approximately, 30% of individuals affected by type 1 DM develop diabetic nephropathy, this increases to 40% in persons affected by type 2 DM [6]. This situation has been compounded by the ever increasing rates of obesity in the world [7]. In 2013 pertaining to the United States of America (USA), the age-adjusted prevalence of obesity was 35% in men and 40% in women representing an upward trajectory of this statistic when juxtaposed to the year 2000 where the overall prevalence of obesity in that country stood at 31% [6, 8]. The connection between obesity and type 2 DM is corroborated by many a research. This only strengthens calls for healthcare professionals involved in the management of this canker to spell out clearly to their clients the gargantuan benefits of lifestyle interventions in the management of DM. Insulin therapy has been a cornerstone in the medical management of type 1 DM. It is also used in the treatment of type 2 DM being necessitated by beta cell exhaustion [9].

This review will attempt to feebly summarize the advances we have made over the years in understanding insulin pertaining to its structure, effects on the various tissues and the development of analogues.

2. An important protein

Insulin's significance to medical science cannot be overemphasized. This small yet mighty peptide hormone together with other counter-regulatory hormones notably glucagon and epinephrine is important in glucose homeostasis in the body. These hormones are important in mediating the switch between anabolic and catabolic phases of the human body enabling it to respond adequately to various stressors such as major burns, sepsis, major surgery and other possibly injurious states. Glucose is the chief fuel used by many cells in the body, typified by neurons and erythrocytes, in cellular respiration to release adenosine triphosphate (ATP) [1, 2].

ATP provides the energy to power life processes of the cell. This is needed to maintain the physiologic milieu of the organism. Too much insulin which is more common in diabetic patients on insulin and insulin secretagogues such as the sulfonylureas and meglitinides may tip the individual into life-threatening hypoglycemia which may be manifested by neurogenic and neuroglycopenic symptoms. Neurogenic symptoms are said to be a response of the autonomic nervous, principally the adrenergic division, to the dip in glucose and are characterized by tremor, anxiety, palpitation, paresthesia, diaphoresis and sensation of hunger. Although, this can be compounded by hypoglycemia unawareness where these symptoms may be blunted and the individual is not able to tell that they are hypoglycemic. Hypoglycemia unawareness affects about 40% of people with type 1 DM. It affects a lesser number of individuals with type 2 DM. Repeated bouts of hypoglycemia lower the level of glucose at which the human body may respond, whiles chronically elevated levels of this may raise this threshold. Neuroglycopenic symptoms which typically occur at lower glucose levels are the result of brain neuronal deprivation of glucose and include confusion, dizziness, headaches, seizures, coma among other symptoms [10, 11].

Perturbation about inducing iatrogenic hypoglycemia presents a hindrance in achieving glycemic targets [10]. Too little insulin for a prolonged period or a reduced action of the hormone may manifest acutely as a decompensated diabetic state such as diabetic ketoacidosis and hyperglycemic hyperosmolar state. Regrettably, the incidence of these two hyperglycemic crises has been on the increase. In the USA, for example, there were 220,340 hospital admissions for DKA in 2017 compared to 168,000 in 2014. Mortality from these conditions though is subsiding. Mortality rate was estimated to be under 1% for DKA, while it can get as high as 20% for HHS. Advanced age, comorbidity and severe dehydration contributed to higher fatality rates in HHS. There is the need for healthcare practitioners and their clients to be more proactive in preventing these dangerous states which can impact a significant economic toll. DKA is the cause of more than 500,000 hospital days per year in the USA according to the Centers for Disease Control and Prevention (CDC). The cost of inpatient care for DKA stood at approximately 6.76 billion US dollars in 2014; in 2017 it was 5.1 billion US dollars [6].

2.1 Discovery

More than a century has elapsed since the crude "isolation" and discovery of insulin by Canadian surgeon Banting and American-Canadian medical student Best with aid from Scottish professor Macleod and Canadian biochemist Collip at the University of Toronto, Canada. Their method was to try to obtain a purified insulin preparation from dog pancreases (later on, rabbits). Their persistence finally paid off in 1921 when their preparation proved capable of normalizing hyperglycemia and glycosuria in dog subjects. That monumental year of 1921 ushered medicine in into a new era. Indeed, insulin has been a pioneer protein for medical research in many regards [12].

In the wake of Paul Langerhans description of islets in the pancreas in 1869, which now eponymously bear his name, scientists had found out that a substance in the pancreas if deficient was implicated in the pathogenesis of diabetes. They had tried to purify this substance from the pancreas for a possible transition from bench-to-beside in the treatment of diabetes. However, despite some strides no one had been able to isolate this substance and demonstrate its reproducible use in man convincingly without toxic reactions up until 1921. In 1889, Minkowksi and von Mering found out that severe diabetes mellitus could be induced in dogs by carrying out a total pancreatectomy leading them to the realization that a substance in the pancreas was needed in the control of blood glucose level. In 1909, Belgian de Meyer suggested the name "insuline" for this substance. Scientific thinking evolved over the years localizing this substance to the islets of Langerhans in the pancreas, thus setting the stage for the work of Dr. Banting and his colleagues. By 1909, Eugene L. Opie had demonstrated histological evidence of hyaline degenerative changes in the islets of Langerhans of individuals with diabetes mellitus [2, 7, 12].

Doctors at Toronto General Hospital led by Walter Campbell were finally able to inject a young boy in 1923 with success to their delight. Fourteen-year old Leonard Thompson showed marked clinical and laboratorial amelioration after he received a second trial of more purified insulin. Testament to its significance, Banting and Macleod received the Nobel Prize in Medicine or Physiology in 1923. Hitherto insulin's discovery, a diagnosis of diabetes mellitus had largely been a death sentence [5].

Of interesting note was the fact that very little was known of insulin's structure at the time of its discovery. To buttress this, Macleod did not think insulin was even a protein. It would take painstaking work of another- Fred Sanger- to elucidate the polypeptide structure of the hormone using 1-fluoro-2,4-dinitrobenzene (DNFB) (Sanger's reagent) among other techniques. He surmised that bovine insulin must be made of two polypeptide chains and deduced their amino acid sequences. Indeed, three different scientific groups set out to confirm Sanger's proposition about insulin's structure with the aid of protein chemical synthesis in the 1960s. They confirmed his earlier findings about insulin's structure by showing that the synthesized protein had biological activity. For his efforts, Sanger received the Nobel Prize much earlier in 1958. X-ray crystallography was used to determine the three-dimensional structure of insulin by Hodgkin, Blundell and Dodson. They published their results in 1969. Yalow and Berson invented radioimmunoassay which proved capable of quantifying insulin levels in the blood in 1959. For this work Yalow received the Nobel prize in 1977, after Berson had died [12].

Again, with the advent of recombinant DNA technology in the 1970s, the gauntlet fell to insulin. It became the second protein after somatostatin to be synthesized by these methods forever revolutionizing the care of individuals with DM. It was the first therapeutic protein engineered by recombinant DNA technology to be approved for human use by the US FDA in 1982. Prior to this the food industry supplied animal pancreases for the extraction of porcine and cattle insulin which both differ from human insulin by 1 (B30) and 3 amino acids (B30, A8 &A10) positions respectively. Recombinant DNA techniques as well as the elucidation of the hormone's structure introduced new paradigms in the treatment of diabetes that allowed mass production of better quality insulin as well as the modification of its chains to produce analogues with desirable properties that more closely mimicked insulin's famous natural biphasic secretion pattern (**Figure 1**) [12, 13].

2.2 Structure and secretion

Insulin is produced by beta cells of the islet of Langerhans in the pancreas. Recent findings suggest that some cells in the brain may also be capable of synthesizing insulin. The constituents of the islets of Langerhans include alpha cells which produce glucagon, beta cells which produce insulin, delta cells which produce somatostatin and pancreatic polypeptide (PP) cells which produce pancreatic polypeptide. Beta cells make up the majority of cells in the islets of Langerhans and are found in the centre of the islets compared to the α cells which tend to be found in the periphery. There human pancreas contains over a million islets of Langerhans. They make up

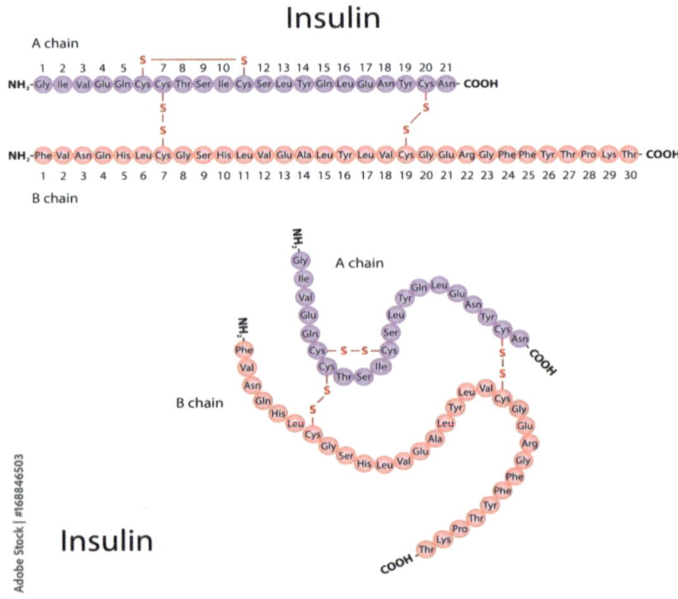

Figure 1.
Primary structure of human insulin.

about 1–2% of the weight of the pancreas, although they receive about 10% of the organ's blood supply. These are scattered groups of endocrine cells within the exocrine acinar cells of the pancreas. The pancreas, just like the liver is both an endocrine and exocrine gland with important roles in digestion and assimilation of food substances. The acinar cells are responsible for manufacture of various enzymes (trypsinogen, chymotrypsinogen, procarboxypeptidase, deoxyribonuclease, ribonuclease, pancreatic lipase and pancreatic amylase) that are important components of the succus entericus which is essential in the hydrolysis of peptides, nucleic acids, lipids and carbohydrates. These are first secreted as zymogens which later undergo activation to function. The exocrine cells in response to secretin produced by enteroendocrine cells in the duodenal mucosa also secrete sodium bicarbonate which serves to neutralize acidic chyme from the stomach among other functions [14, 15].

Insulin is a peptide hormone made up of 51 amino acids, with a molecular weight of 5802 and its isoelectric point at pH 5.5. It is composed of two chains: a B chain with 30 amino acids and an A chain with 21 amino acids. These chains are joined together by two interchain disulphide bonds connected via cysteine residues. The A chain also has an intrachain disulphide bond. The gene for insulin's synthesis is located on chromosome 11p [2].

Insulin is first synthesized as 110-amino acid chain known as preproinsulin by polysomes. Preproinsulin consists of an N-terminal signal peptide, a B chain, a connecting peptide (C-peptide) and a carboxyl-terminal A chain. The N-terminal signal peptide directs insulin to the endoplasmic reticulum where it is cleaved off by an endopeptidase. The protein undergoes further repackaging in the Golgi bodies and is transmitted through the trans-Golgi network. It is incorporated into secretory vesicles from which it can be released from beta cells by exocytosis for example upon glucose stimulation [16]. Glucose is transported into human beta cells by glucose transporters (GLUT) 1 and 3; GLUT2 is currently thought to be more important in this function in rodents. Although, the question about the functions of GLUTs in the human β cell is still not completely answered. Upon cell entry, glucose is phosphorylated by glucokinase to form glucose 6-phosphate [1]. Eventually, ATP is formed by oxidative processes in the cell. ATP-sensitive K+ channels in the cell membrane of the beta cell subsequently close as the ratio of ATP/ADP molecules in the β cell increases. This causes depolarization of the cell and resultant movement of Ca^{2+} ions into the cell via voltage dependent Ca^{2+} channels. These ions are important in causing glucose stimulated insulin secretion (GSIS) into the bloodstream. The mechanism of action of sulphonylureas which are a drug class used in used in the treatment of type 2 DM also involves closure of ATP-sensitive K^+ channels independent of blood glucose levels by their binding to the sulphonylurea receptor of these particular channels. GSIS occurs by both ATP-regulated K^+dependent and ATP-regulated K^+ independent pathways. Insulin secretion in response to glucose is classically said to be biphasic. In that there is an initial rapid phase lasting a few minutes in response to postprandial hyperglycemia, then a more sustained second phase. The increase in intracellular calcium ions causes exocytosis of secretory granules in the readily realizable pool (RRP). Typically, blood glucose concentrations above 5 mmol/l are necessary to cause the initial surge in insulin secretion. Lower levels of between 2 and 4 mmol/l, however are enough to cause synthesis of insulin into secretory granules. This is necessary to ensure that β cells can respond rapidly to the body's metabolic demands [1, 2]. Insulin exists as a hexamer complexed with zinc ions in the secretory granules.

Nutrient in the gastrointestinal tract are able to cause insulin secretion which is not dependent on glucose. The incretin hormones, that is gastric inhibitory polypeptide

(GIP) and glucagon-like peptide 1 (GLP1), are implicated in this process. While nutrients and non-nutrients can cause the initial phase, the more sustained phase is mediated primarily by nutrients. Amino acids like alanine cause an increase in insulin secretion. Non-nutrient secretagogues are also able to cause an increase in insulin secretion. There are endocrine, autocrine, paracrine and neural factors that affect insulin release. Somatostatin inhibits both insulin and glucagon secretion. While insulin downregulates glucagon secretion, glucagon upregulates insulin secretion. Insulin is thought to have positive effects on its own secretion. Cholinergic stimulation increases insulin secretion, while activation of α2 receptors by sympathetic nervous system inhibits insulin secretion. Vasoactive intestinal peptide from peptidergic fibers also cause an increase insulin secretion [1, 2].

2.3 Insulin's effect on tissues and mechanism of action

The insulin receptor (IR) responsible for mediating insulin's effects is a tyrosine kinase with a heterotetrameric structure consisting of 2 alpha and 2 beta subunits and is located on the cell surface membrane. The gene for the insulin receptor is 150-kb long, has 22 exons and is on human chromosome 19p13.3–p13.2. Insulin binds to the extracellular alpha units, with the intracellular part of the beta subunit involved in recruitment of several adaptor proteins including the insulin receptor substrates (IRS) which mediate insulin's intracellular actions via phosphorylation. There are two sites in the extracellular portion of the receptor with different affinities where insulin is capable of binding to. The IR is said to exhibit negative cooperative binding. There are two known isoforms of the receptor: IRA and IRB. IRA has fewer amino acids due to alternative splicing of exon 11. IRB is more specific for insulin than IRA. Three regions in the beta subunit of note undergo trans-autophosphorylation: one in the juxtamembrane domain (Y_{965} and Y_{972}), one in the activation loop (Y_{1162}, Y_{1158} and Y_{1163}) and another in the C-terminal portion of the receptor (Y_{1328} and Y_{1334}) after insulin binds to the extracellular portion to cause a transformational change in shape. IRS 1 and 2 are more widely distributed in the tissues. Upon activation of the IR, after phosphorylation of NPEpY732 in the juxtamembrane zone of the receptor, IRS-1 is recruited to this site. IRS-1 has amino-terminal pleckstrin homology (PH) and phosphotyrosine binding (PTB) regions which are necessary for membrane and receptor association. Also, it has an unstructured part consisting of 14 tyrosine phosphorylation sites which follow these two other regions and undergo phosphorylation when the IRS-1 is activated. After phosphorylation, two YMPM motifs in IRS-1 bind src-homology 2 (SH2) binding sites of p85α which is a regulatory subunit of phosphatidylinositol-3-kinase (PI3K). The catalytic portion PI3K then becomes able to phosphorylate phosphatidylinositol 4,5-bisphosphate (PIP2) to phosphatidylinositol 3,4,5-triphosphate (PIP3) which is responsible for some of insulin's downstream actions including the activation of phosphoinositide – dependent protein kinase 1 (PDK-1). PDK-1 activates AKT. AKT2 mainly mediates insulin's metabolic functions in skeletal muscle, adipose tissue, hepatic tissue and other bodily tissues. These processes will be discussed in more elaborate detail in the text below. It is now known that insulin, as well as the insulin-like growth factors (IGFs) 1 and 2, has important roles in other tissues of the body apart from the striated muscle, fat and the liver mediated by various receptors [1, 2, 7, 17].

Insulin just like IGF-1 has mitogenic actions. Insulin plays various roles in cell growth, division, differentiation and survival mainly via mitogen activated protein kinase pathway (MAPK). The insulin concentration needed for its mitogenic functions is greater than that for its metabolic activities. This involves the binding of IRS-1 to growth factor receptor-bound protein 2 (Grb2) which occurs after insulin signaling via the insulin receptor. Son of sevenless Ras/Rac guanine nucleotide exchange factor 1 (SOS1) is recruited by Grb2 leading to the activation of the rat sarcoma protein (RAS) via GTP-GDP exchange. RAS then involves c-Raf which phosphorylates MAPK/extracellular-signal regulated kinase (MERK), which then activates extracellular-signal regulated kinase (Erk). Translocation of Erk to the nucleus leads to the transcription of nuclear factors important in mitogenesis [1, 2, 17].

2.3.1 Skeletal muscle

The skeletal muscle is the primary site of postprandial glucose uptake. Insulin acting on the IR causes the translocation of GLUT4 receptor to the surface membrane. This effect is mediated via the AKT pathway. This involves the phosphorylation of Cbl-associated protein (CAP) and CAP: CBL: CRKII complex formation. GLUT4 shuttles glucose molecules into the cell by facilitated diffusion. The translocation of GLUT4 is reversible as insulin concentrations dwindle. Insulin is thought to also influence GLUT4 gene expression in muscle and adipose tissue. Once inside the cell, glucose can be used for energy release or stored as glycogen. Insulin upregulates glycogenesis n skeletal muscle and inhibits glycogenolysis [1, 17, 18].

2.3.2 Adipose tissue

Insulin stimulates white adipose tissue to undergo lipogenesis and inhibits lipolysis. Glucose is converted to triacylglycerol in the fed state. The functions of adipose tissue, especially visceral fat is of special interest in the pathogenesis of hyperinsulinemia, insulin resistance, obesity and type 2 DM. Insulin resistance is when the response of target cells to insulin is lower than it should be. In cases of nutrient excess, when the storage capacity of fat tissue is exceeded, lipases hydrolyse triacylglycerol into free fatty acids (FFA) and glycerol which enter the circulation. Insulin is unable to inhibit lipolysis, and this may lead to "lipotoxicity". Ectopic accumulation of lipids occurs in skeletal muscle and hepatocytes. This encourages the development of non-alcoholic fatty liver disease (NAFLD). In skeletal muscle, this intramyocellular accumulation sometimes is associated with the development insulin resistance [1, 17, 18].

2.3.3 Liver

The liver has a special role in the regulation of glucose homeostasis. The liver is capable of converting non-carbohydate sources like alanine, lactate and pyruvate into glucose in a process known as gluconeogenesis. This is upregulated by glucagon and inhibited by insulin and alcohol. Insulin also inhibits glycogenolysis but upregulates glycogenesis. The liver typically holds about 100 g of glycogen, with about 500 g in skeletal muscle. This helps to explain partly the tendency of patients with liver cirrhosis to suffer hypoglycemia. GLUT 2 receptor is bidirectional as it is involved in both glucose uptake and release into the blood [1, 7, 18].

When there is lipotoxicity, hyperinsulinemia is unable to suppress the release of non-esterified fatty acids (NEFA) by adipose tissue. The liver uses some of these excess fatty acids to synthesize VLDL leading to hypertriglyceridemia. Also, individual are prone to suffering non-alcoholic steatohepatosis (NASH) and non-alcoholic fatty liver disease (NAFLD). Infection with hepatitis C virus is thought to also potentiate such pathways. Insulin-resistance may lead to uncontrolled gluconeogenesis with attendant chronic hyperglycemia precipitating end-organ damage. Dysregulation of forkhead box O family (FOXO) of transcription factors by defective insulin PI3K/AKT signaling allows the unrestrained translocation of these factors to the nucleus to promote gluconeogenesis [17, 18].

2.3.4 Brain

The brain is selective for glucose as its principal fuel, though it is capable of switching in lean times. GLUT1 receptors at the blood-brain barrier are capable of extracting glucose from the blood in an insulin-independent fashion. Recent findings though suggest that certain areas in the brain such as the pineal gland choroid plexus and the pituitary as well as the spinal cord obtain glucose in an insulin-dependent fashion. Studies show that insulin may improve cognition and memory in individuals with Alzheimer's disease (AD). Besides this, patients with AD are more prone to developing type 2 DM. This link has generated much excitement in medical circles with some practitioners labeling AD as "type 3 diabetes". Intranasal insulin is being tried in this respect [1, 19].

2.3.5 Endothelial tissue

Insulin acts via the insulin receptor on endothelial tissues. Aberrant insulin signaling associated with selective insulin resistance may lead reduced nitric oxide production, endothelial dysfunction, poor wound-healing, atherogenesis, pro-thrombotic and pro-inflammatory states with an increased risk of hypertension and other cardiovascular morbidity and mortality. Novel targeted therapies are needed to attenuate these effects [1].

2.4 Insulin today

When insulin was first made available clinically, large volumes of the product had to be given which often led to local site infections and abscesses. These problems have reduced for today's diabetic patient. Also, increased ease of delivery of insulin and glucose monitoring have improved the quality of life of people living with diabetes. Infusion pumps and insulin pens have allowed more efficient drug administration. Blood glucose monitoring with capillary glucose tests, continuous glucose monitoring and widespread availability of HbA1C test have allowed more precise glycemic control and attenuated the risks of diabetic complications [9].

Today's insulin's preparations have been upgraded with refinements in their peak action times and duration of action, although they are still a poor substitute for insulin's complex physiology with its feedback systems. They have facilitated the use of today's basal bolus regimen in the management of diabetes mellitus.

There are rapid-acting, short-acting, intermediate-acting and long-acting insulin analogs. Rapid-acting insulin analogs include lispro, aspart and glulisine which have on the average an onset of action of about 5–15 min, a usual peak action time of about 30–60 min and duration of 2–5 h. Short-acting insulin analogs such as regular human insulin typically have an onset of action of time of about 30–60 min, a usual peak action time of about 1–3 h and duration of 4–8 h. Intermediate-acting insulin analogs include Neutral Protamine Hagedorn (NPH) on the average have an onset of action of time about 1–2 h, a usual peak action time of about 4–8 h and duration of 8–12 h. Long-acting analogs including detemir, and glargine have on the average an onset of action of time of about 30–60 min, with no peak action time, last about 16–24 h and can generally be dosed as a single shot per day. There are also premixed analog formulations available which combine different types of analogs [9, 20]. Concerns have been raised about the theoretical risk of analogs to cause cancer through insulin's mitogenic pathways. Evidence to support this stance though remains paltry, and more research work will need to be done in this regard [21]. Economic barriers remain in acquiring these novel therapies perhaps in ironic contrast to this saying attributed to Banting, "insulin does not belong to me, it belongs to the world [5]."

3. Conclusion

It goes without saying that all healthcare professionals especially doctors should be abreast of the ever-changing landscape in the management of DM. The pervasive nature of this malady in surgery, internal medicine, child health and obstetrics & gynecology make this imperative. This of course will be a nine-day wonder without an understanding of insulin's mechanisms not only in type 1 DM, but also in other forms of DM. This in the long run will boost patient outcomes. Medical research must also continue with necessary funding. Additionally, it will be of great benefit for pharmaceutical companies to consider the costs of drugs so as to make them accessible to society's most vulnerable. As history has shown us, it is only on the intrepid steps of alacrity and sometimes good old serendipity will other scientific breakthroughs be made in the treatment of DM. As the torch is passed on to this generation of medical practitioners in the next century after insulin's discovery we owe a lot to the prescience of the men who have contributed to our understanding of this interesting peptide hormone. It is only befitting tribute to their memories that we expand their work and finally "defeat" DM together.

Conflict of interest

The author declares no conflict of interest.

Author details

Michael Awuku
St. Michael's Hospital, Pramso, Ghana

*Address all correspondence to: unclefii07@gmail.com

References

[1] Rahman MS, Hossain KS, Das S, Kundu S, Adegoke EO, Rahman MA, et al. Role of insulin in health and disease: An update. International Journal of Molecular Sciences. 2021;**22**:6403. DOI: 10.3390/ijms22126403

[2] Wilcox G. Insulin and insulin resistance. Clinical Biochemist Reviews. 26 May 2005;**26**:19-23

[3] Alicic RZ, Rooney MT, Tuttle KR. Diabetic kidney disease: Challenges, Progress, and possibilities. Clinical Journal of the American Society of Nephrology. 2017;**12**(12):2032-2045. DOI: 10.2215/CJN.11491116: 10.2215/CJN.11491116

[4] Forst T, Mathieu C, Giorgino F, Wheeler DC, Papanas N, Schmieder RE, et al. New strategies to improve clinical outcomes for diabetic kidney disease. BMC Medicine. 2022;**20**:337. DOI: 10.1186/s12916-022-02539-2

[5] Lewis GF, Brubaker PL. The discovery of insulin revisited: Lessons for the modern era. The Journal of Clinical Investigation. 2021;**131**(1):e142239. DOI: 10.1172/JCI142239

[6] Gosmanov AR, Gosmanova EO, Kitabchi AE. Hyperglycemic Crises: Diabetic Ketoacidosis and Hyperglycemic Hyperosmolar. Endotext. South Dartmouth (MA): MDText.com, Inc.; 2021, 2000

[7] Petersen MC, Shulman GI. Mechanisms of insulin action and insulin resistance. Physiological Reviews. 2018;**98**:2133-2223. DOI: 10.1152/physrev.00063.2017

[8] Bentley RA, Ormerod P, Ruck DJ. Recent origin and evolution of obesity-income correlation across the United States. Palgrave Communications. 2018;**4**:146. DOI: 10.1057/s41599-018-0201-x

[9] Bolli GB, Cheng AYY, Owens DR. Insulin: Evolution of insulin formulations and their application in clinical practice over 100 years. Acta Diabetologica. 2022;**59**:1129-1144. DOI: 10.1007/s00592-022-01938-4

[10] Nakhleh A, Shehadeh N. Hypoglycemia in diabetes: An update on pathophysiology, treatment, and prevention. World Journal of Diabetes. 2021;**12**(12):2036-2049. DOI: 10.4239/wjd.v12.i12.2036

[11] Martín-Timón I, del Cañizo-Gómez FJ. Mechanisms of hypoglycemia unawareness and implications in diabetic patients. World Journal of Diabetes. 2015;**6**(7):912-926. DOI: 10.4239/wjd.v6.i7.912

[12] Flier JS, Ronald Kahn C. Insulin: A Pacesetter for the Shape of Modern Biomedical Science and the Nobel Prize. München, Germany: Elsevier GmbH; 2021. DOI: 10.1016/j.molmet.2021.101194

[13] Riggs AD. Making, cloning, and the expression of human insulin genes in bacteria: The path to Humulin. Endocrine Reviews. 2021;**42**(3):374-380. DOI: 10.1210/endrev/bnaa029: 10.1210/endrev/bnaa029

[14] Bilous R, Donelly R. Handbook of Diabetes. 4th ed. Chichester, West Sussex, UK: Wiley-Blackwell; 2010. p. 24

[15] Victor P. Eroschenko. difiore's Atlas of Histology with Functional Correlations. 12th ed. Philadelphia, PA, USA: Lippincott Williams & Wilkins; 2013. p. 378

[16] Vasiljević J, Torkko JM, Knoch K-P, Solimena M. The making of insulin

in health and disease. Diabetologia. 2020;**63**:1981-1989. DOI: 10.1007/s00125-020-05192-7

[17] White MF, Ronald Kahn C. Insulin Action at a Molecular Level - 100 Years of Progress. München, Germany: Elsevier GmbH; 2021. DOI: 10.1016/j.molmet.2021.101304

[18] Czech MP. Insulin action and resistance in obesity and type 2 diabetes. Nature Medicine. 2017;**23**(7):804-814. DOI: 10.1038/nm.4350

[19] Walker JM, Harrison FE. Shared neuropathological characteristics of obesity, type 2 diabetes and Alzheimer's disease: Impacts on cognitive decline. Nutrients. 2015;7:7332-7357. DOI: 10.3390/nu7095341

[20] Shukla Gupta S, Acharya S, Shukla S. A look into the next century after 100 years of insulin. Cureus. 2022;**14**(10):e30133. DOI: 10.7759/cureus.30133

[21] Sciacca L, Le Moli R, Vigneri R. Insulin analogs and cancer. Frontiers. Feb 2012:1-4. DOI: 10.3389/fendo.2012.00021

Chapter 2

Type 1 Diabetes and Beta Cells

Sheila Owens-Collins

Abstract

This book chapter provides an overview of Type 1 diabetes, focusing on the role of beta cells, autoimmunity, genetics, environmental factors, and beta cell health. While genetic factors are also important, environmental factors such as viral infections and dietary factors may trigger or accelerate the development of Type 1 diabetes. Maintaining beta cell health is essential for the prevention and management of Type 1 diabetes. Factors such as glucose toxicity, oxidative stress, and inflammation can contribute to beta cell dysfunction and death. The chapter discusses transplantation of islet cells both primary and stem cell-derived to treat diabetes. The chapter also outlines the stages of Type 1 diabetes development, starting with the pre-symptomatic stage and progressing to the onset of symptoms, the clinical diagnosis, and the eventual need for insulin therapy. Supporting hormones, such as insulin, glucagon, amylin, somatostatin, and incretin hormones, play critical roles in maintaining glucose homeostasis. Finally, the chapter highlights the effect of food on beta cell health and the effect of various drugs and medications used to manage diabetes.

Keywords: beta cells, diabetes, insulin, autoimmunity, diabetes medications

1. Introduction

Type 1 diabetes (T1D) is a chronic condition in which the body's immune system attacks and destroys the insulin-producing cells in the pancreas, called beta cells. Insulin is a hormone that regulates the amount of glucose (sugar) in the bloodstream and allows the body to use it for energy. Without enough insulin, glucose builds up in the blood and can lead to a range of health problems.

T1D is usually diagnosed in children and young adults, although it can occur at any age. The exact cause of T1D is not known, but genetic and environmental factors are thought to play a role. Symptoms of T1D can include frequent urination, increased thirst, extreme hunger, unexplained weight loss, fatigue, and blurred vision.

There is no cure for T1D, but with proper treatment and management, people with the condition can live long and healthy lives. Ongoing research is focused on improving treatments and finding a cure for this chronic disease. This book chapter aims to highlight key aspects of diabetes and beta cells.

2. Autoimmunity and diabetes

Genetics can play a role in the development of type 1 and type 2 diabetes [1–4]. In the case of T1D, genetic factors can contribute to the destruction of beta cells in the pancreas, which leads to a lack of insulin production. This destruction is often caused by an autoimmune response, where the body's immune system attacks and destroys the beta cells. The exact cause of this autoimmune response is not fully understood, but both genetic and environmental factors are thought to play a role.

Several genetic factors have been identified that can contribute to the destruction of beta cells in the pancreas and the development of T1D [2]. Some of these factors include:

2.1 Genetic factors

Certain genetic variations have been associated with an increased risk of developing T1D. These variations affect the immune system and can make individuals more susceptible to autoimmune diseases.

Certain variations in the human leukocyte antigen (HLA) genes have been associated with an increased risk of developing T1D [5]. These genes play a key role in regulating the immune system and can affect how the body recognizes and attacks foreign invaders, including beta cells.

Variations in the insulin gene (INS) have also been associated with an increased risk of developing T1D [6]. This gene plays a role in the production of insulin, and variations can affect the development and function of beta cells [7–10].

The protein tyrosine phosphatase non-receptor type 22 (PTPN22) gene is involved in regulating the immune system and has been associated with an increased risk of developing T1D [11–13]. The Trp620 variant of PTPN22 gene is also associated with other autoimmune disorders [14–16].

The cytotoxic T-lymphocyte antigen 4 (CTLA4) gene encodes a molecule that functions as a negative regulator of T-cell activation. Polymorphisms in this gene are involved in regulating the immune system and has been associated with an increased risk of developing T1D [17–20].

Other genes involved in immune system regulation, such as IL2RA/CD25 [21, 22], IL7RA, SUMO4 [23, 24] and IFIH1 [25, 26], have also been associated with an increased risk of developing T1D.

Given the autoimmune basis of T1D autologous hematopoietic stem cell transplantation (auto-HSCT), a type of auto-transplantation where hematopoietic stem cells are removed from a patient, enriched and introduced back into the patient [27–30] could regenerate immune tolerance against auto-antigens and be associated with lasting and complete remission of T1D.

2.2 Environmental factors

Environmental factors such as viral infections, toxins, and diet have also been linked to the development of T1D [31]. Some researchers believe that viral infections can trigger the immune system to attack the beta cells, leading to the development of T1D. In particular, enteroviruses have been linked to an increased risk of developing T1D [32, 33]. Exposure to toxins such as nitrosamines [34] and bisphenol A (BPA) [35] has also been associated with an increased risk of developing T1D.

2.3 Other factors

Other factors that may play a role in the development of autoimmune diabetes include the gut microbiome [36, 37], stress [38], and low levels of vitamin D [39].

Overall, the development of autoimmune diabetes is likely due to a combination of genetic and environmental factors. In the case of type 2 diabetes, genetics can also play a role in the development of the condition [40, 41]. Certain genes can affect how the body processes insulin, leading to insulin resistance and an increased risk of developing type 2 diabetes. However, lifestyle factors such as diet, exercise [42] and stress [38] also play a significant role in the development of type 2 diabetes.

3. Beta cell health

The classic view of immune response gone wrong in T1D is that autoreactive T cells mistakenly destroy healthy β-cells. There is an alternative view that in response to cell stress an unhealthy β-cell provokes an immune attack that negatively effects the source of insulin [43, 44]. Restoring β-cell health therefore is proposed as an important component of treating T1D. Beta cell health can be affected by a variety of factors, including:

3.1 Glucose and lipid levels

Chronic exposure to high levels of glucose and lipids in the blood can lead to beta cell dysfunction and damage, impairing their ability to produce and release insulin effectively. The beta cell workload hypothesis [45] is a theory that suggests that chronic exposure of beta cells in the pancreas to high levels of glucose and lipids (fatty acids) can lead to beta cell dysfunction and the development of type 2 diabetes.

3.2 Oxidative stress

Beta cells are known to express lower levels of antioxidant enzymes like catalase and glutathione peroxidase that are required to protect against reactive oxygen species (ROS), thus making them have higher susceptibility to ROS damage [46]. This oxidative stress can damage cell membranes, DNA, and other cellular components, impairing beta cell function [47, 48]. Oxidative stress leads to loss of β-cell identity by downregulation of maturity genes like MAFA and PDX1, and upregulation of progenitor genes like SOX9 and HES1 [49, 50]. This makes beta cells less beta-cell like and leads to further decrease in insulin secretion.

3.3 Age

Beta cell function tends to decline with age, contributing to an increased risk of developing diabetes in older individuals [51, 52]. This decline in beta cell function may be due to a combination of factors, including increased oxidative stress, inflammation, and mitochondrial dysfunction, which can accumulate over time and impair beta cell survival and function.

In addition to these cellular changes, aging is also associated with changes in hormonal and metabolic regulation, which can further contribute to beta cell dysfunction. For example, age-related changes in insulin sensitivity and glucose metabolism can increase the demand on beta cells, leading to increased oxidative stress and cellular damage. Furthermore, changes in hormones such as growth hormone and cortisol, which occur with aging, can impair beta cell function and contribute to the development of diabetes.

3.4 Other factors

Inflammation and genetic factors: Chronic inflammation can also contribute to beta cell dysfunction, damage [53] and dedifferentiation [54], as well as impairing insulin sensitivity [55]. Certain genetic variations can affect beta cell function and increase the risk of developing diabetes. These factors have been discussed in detail in Section 2.1.

Obesity and physical activity: Excess body fat, particularly abdominal fat, can contribute to beta cell dysfunction and insulin resistance, increasing the risk of developing diabetes [56, 57]. Regular physical activity can improve beta cell function and insulin sensitivity, reducing the risk of developing diabetes [58].

Medications and environmental toxins: Certain medications, such as corticosteroids, can impair beta cell function and increase the risk of developing diabetes. These effects have been discussed elsewhere in this chapter. Exposure to environmental toxins, such as bisphenol A (BPA) and phthalates, can also impair beta cell function and increase the risk of developing diabetes. These effects have been discussed in Section 2.2.

Overall, maintaining beta cell health is important for preventing and managing diabetes, and a healthy lifestyle that includes regular physical activity, a balanced diet, and avoidance of environmental toxins can help support beta cell function.

4. Islet cell transplants

4.1 Islet cell auto-transplantation

Auto-transplantation is a type of islet transplantation where the islet cells are taken from the patient's own pancreas and transplanted back into the patient's liver [59]. This procedure is typically used in cases where the patient's pancreas has been removed or damaged due to disease or injury. Auto transplantation may also be used to treat chronic pancreatitis [60], a condition that causes inflammation and scarring of the pancreas, which can lead to diabetes.

Stem Cell Derived Beta Cell (SC-β cell) transplantation is a special type of auto-transplantation where β cells derived from a patient's stem cells can be transplanted into the patient [61, 62]. SC-β cell therapy is still in its early stage where work is being done to understand how to prevent cell death post engraftment [63, 64].

4.2 Islet cell allo-transplantation

Allotransplantation is a type of islet transplantation where the islet cells are taken from a donor pancreas and transplanted into the recipient's liver [59]. This procedure is typically used in patients with T1D who have severe hypoglycemia unawareness or difficult-to-control blood sugar levels despite optimal medical therapy. Allotransplantation can provide long-term insulin independence for some patients, but it requires immunosuppressive medications to prevent rejection of the transplanted cells.

Both auto-transplantation and allotransplantation have advantages and disadvantages. Auto-transplantation avoids the need for immunosuppressive medications, as the transplanted cells are from the patient's own body, but it may not be effective for all patients with diabetes. Allo-transplantation can provide long-term insulin independence, but it requires immunosuppressive medications, which can have side effects and increase the risk of infections and other complications.

In summary, both auto-transplantation and allo-transplantation of islet cells are potential strategies to treat diabetes. However, their use depends on the individual patient's medical history and condition.

5. Stages of type 1 diabetes development

JDRF (formerly known as the Juvenile Diabetes Research Foundation) has identified several stages of T1D development, based on the progression of autoimmunity and beta cell destruction [65]. These stages are:

Stage 1—Autoimmunity: During this stage, the immune system begins to attack the beta cells in the pancreas, but there are no symptoms of diabetes yet. This stage can last for months or even years.

Stage 2—Abnormal blood sugar levels: During this stage, the destruction of beta cells has progressed to the point where blood sugar levels begin to rise, but there are still no symptoms of diabetes. This stage is also known as preclinical T1D.

Stage 3—Clinical onset of diabetes: During this stage, the destruction of beta cells has reached a critical threshold, and the individual begins to experience symptoms of diabetes, such as frequent urination, excessive thirst, and weight loss. At this point, the individual is diagnosed with T1D.

6. Supporting hormones for glucose homeostasis

There are supporting hormones that complement insulin's role in regulating glucose levels in the body. Glucagon, amylin, somatostatin, and incretin hormones are all involved in regulating glucose homeostasis in the body [66]. Here are brief summaries of their roles:

Glucagon is produced by the alpha cells of the pancreas in response to low blood glucose levels. Its main function is to stimulate the production and release of glucose from the liver, increasing blood glucose levels [67].

Amylin is co-secreted with insulin from the beta cells of the pancreas. Its main function is to slow down gastric emptying and suppress glucagon secretion, which helps to regulate blood glucose levels [68].

Somatostatin is produced by the delta cells of the pancreas and inhibits the release of insulin, glucagon, and other hormones that affect glucose metabolism [69].

Incretin hormones, such as glucagon-like peptide-1 (GLP-1) and gastric inhibitory polypeptide (GIP), are produced by the intestines in response to food intake. They stimulate insulin secretion from the pancreas, inhibit glucagon secretion, and slow down gastric emptying, all of which help to regulate blood glucose levels [70].

Overall, the actions of these hormones work together to maintain glucose homeostasis in the body. Insulin and amylin work to lower blood glucose levels, while glucagon and incretin hormones work to increase or decrease blood glucose levels depending on the body's needs. Somatostatin plays a regulatory role by inhibiting the secretion of other hormones to maintain balance. Dysfunction in any of these hormones can lead to impaired glucose regulation, which can contribute to the development of diabetes.

7. Infections associated with diabetes

People with diabetes are generally more susceptible to infections due to a weakened immune system [71]. Some infections that are particularly associated with diabetes include:

7.1 Urinary tract infections (UTIs)

There are several reasons why people with diabetes are more susceptible to urinary tract infections (UTIs). High blood glucose levels can cause bacteria to grow in the urinary tract, leading to UTIs [72]. Diabetes can damage nerves that control bladder function, leading to incomplete bladder emptying and increased urine retention, which can also increase the risk of UTIs [73, 74]. Diabetic patients also tend to have increased adherence of bacteria to uroepithelial cells [75]. There is a downregulation of the antimicrobial peptide psoriasin and this increases bacterial burden in the urinary bladder [76]. This too may play a role in the pathogenesis of UTIs in diabetics.

7.2 Skin infections

High blood glucose levels can lead to dry, cracked skin, making it easier for bacteria to enter and cause infections. High blood glucose levels can also promote the growth of certain types of bacteria, leading to an overgrowth of harmful bacteria on the skin. Poor wound healing and decreased immunity can also contribute to the development and persistence of skin infections in people with diabetes. Common skin infections in people with diabetes include cellulitis, styes, and boils [77].

7.3 Fungal infections

High blood glucose levels can also create a favorable environment for fungal infections, particularly in the mouth, genital area, and feet. Yeasts like Candida spp. can utilize glucose as a viable nutrient [78]. Since diabetics with uncontrolled hyperglycemia have increased glucose in their secretions, they are at a higher risk for fungal infection. Diabetes can create an imbalance in the body's natural microbiota, which can promote fungal growth [79]. Common fungal infections in people with diabetes include oral thrush, vaginal yeast infections, and athlete's foot [80].

7.4 Respiratory infections

People with diabetes may be more susceptible to respiratory infections, such as pneumonia and bronchitis, due to a weakened immune system [81]. Diabetes can damage the small blood vessels and nerves in the lungs [82], making it more difficult to clear mucus and other secretions from the airways. This can create an environment that is more conducive to bacterial growth and increase the risk of respiratory infections.

7.5 Tuberculosis (TB)

People with diabetes are at increased risk of developing TB, a bacterial infection that mainly affects the lungs [83] due to several factors. The weakening of the immune system makes it harder for the body to fight off infections such as TB. This can allow the TB bacteria to take hold and multiply, leading to an active TB infection. Diabetes can also increase the risk of latent TB infection (LTBI), which occurs when a person is infected with TB bacteria but does not develop active TB disease [84, 85]. People with diabetes are more likely to progress from LTBI to active TB disease, as diabetes weakens the immune system's ability to control the TB bacteria. Finally, people with diabetes may have other risk factors for TB, such as malnutrition, poverty, and overcrowding, which can increase their risk of exposure to TB bacteria [86].

It's important for people with diabetes to practice good hygiene and take steps to prevent infections, such as keeping blood glucose levels under control, washing hands frequently, and getting vaccinated against preventable infections like the flu and pneumonia.

8. Foods affecting β-cells

In this section we will cover vitamins, supplements, and foods that improve or impair β-cell function and ultimate blood glucose control [87]. Consuming too much carbohydrates, especially those with a high glycemic index increases stresses on β-cells and risk of diabetes. The key to preventing insulin resistance and protecting β-cells is to increase intake of foods that allow blood sugar to rise slowly. Dietary fiber slows digestion and release of sugars and that is key to reduce blood sugar spikes. Simple carbohydrates like the sugars found in candy, syrups, etc. are easily digestible causing rapid increase in blood sugar levels and stressing the β-cells to rapidly produce large amounts of insulin.

Vitamin deficiencies have been linked with increased risk for diabetes mellitus.

8.1 Vitamin A (VA)

Vitamin A (VA) is the name of a group of fat-soluble retinoids [88]. In animal models, VA deficiency has been linked with β-cell death, insulin insufficiency, and hyperglycemia [89]. While incorporation of VA in the diet has been shown to improve hyperglycemia and glucose tolerance [90]. Islet stellate cells (ISCs) are VA-storing cells in pancreatic islets. ISCs when activated are implicated in islet fibrosis which reduces β-cell mass and glucose tolerance [91]. VA deficiency reduces islet function by activating ISCs, while reintroduction of dietary VA can restore pancreatic VA levels, endocrine hormone profiles, and inhibit ISCs activation [92].

8.2 Vitamin B6 (VB6)

Vitamin B6 is a cofactor in various metabolic reactions that regulate glucose, lipids, and amino acids [93]. Because of this VB6 deficiency impairs glucose and lipid metabolism. Mutations in genes involved in vitamin B6 metabolism cause diabetes [94, 95]. Reduced VB6 availability affects T-cell composition [96], which may contribute to pancreatic islet autoimmunity in T1D.

8.3 Vitamin B9 or folate (VB9)

Depletion of VB9 causes oxidative stress, abnormal glucose and lipid metabolism, insulin resistance, and endothelial disruption [97]. Folate supplementation lowers insulin resistance and improves glucose metabolism [98, 99]. While the exact mechanisms linking folate deficiency and beta cell health are not yet fully understood, it is thought that folate may play a role in regulating the expression of genes involved in beta cell function and survival [100].

8.4 Vitamin B12 (VB12)

Vitamin B12 is a cobalt containing vitamin and is therefore also called as cobalamin [101]. VB12 has many physiological functions but its effect on diabetes comes from its role as a cofactor for methionine synthase which catalyzes the conversion of homocysteine to the essential amino acid methionine [102]. Homocysteine promotes oxidative stress, autoimmunity, insulin resistance, β-cell dysfunction, systemic inflammation, obesity, and endothelial dysfunction. Reduced VB12 availability affects homocysteine levels and through that route promote β-cell dysfunction [103–105].

8.5 Vitamin C or ascorbic acid (VC)

Vitamin C cannot be synthesized by humans and has to be obtained from diet [106]. Ascorbic acid acts as a cofactor for 15 mammalian enzymes [107]. Some studies suggest that VC deficiency predisposes to T2DM [108] and VC supplementation reduced fasting glucose levels in patients with T2DM [109]. VC administration has been shown to reduce blood glucose and increase superoxide dismutase and glutathione levels, resulting in reduced insulin resistance by lowering oxidative stress [110]. VC prevents sorbitol accumulation and glycosylation of proteins and thus reduces the microvascular complications of diabetes [111]. Overall, while the relationship between ascorbic acid deficiency and beta cell health is still not fully understood, there is evidence to suggest that optimizing ascorbic acid status through a healthy diet or supplementation may be beneficial for preserving beta cell function and reducing the risk of diabetes or its progression [112].

8.6 Vitamin D or calciferol (VD)

Vitamin D is a fat soluble vitamin that is obtained through diet and endogenously when ultraviolet (UV) rays from sunlight strike the skin and trigger VD synthesis [113]. VD deficiency may promote β-cell autoimmunity [97, 114], cause insulin resistance, insulin insensitivity, and impaired insulin secretion through β-cell dysfunction [115–119]. VD has been shown to prevent epigenetic alterations associated with insulin resistance and diabetes [120]. Some clinical studies [121–124] have shown that calciferol supplementation was associated with the improvement of insulin secretion while others have not shown a statistically significant benefit [125–127]. Overall, while the relationship between VD deficiency and beta cell health is still not fully understood, optimizing calciferol status through a healthy diet or supplementation may be beneficial for preserving beta cell function and reducing the risk of T1D.

8.7 Vitamin E (VE)

Vitamin E is the collective name for a group of eight fat-soluble compounds that have distinctive antioxidant activities [128]. As a dietary anti-oxidant VE inhibits lipid per oxidation [129]. Significant correlation has been observed between the increased blood sugar levels and the depletion of the antioxidants [130] and higher dietary anti-oxidant capacity is inversely associated with prediabetes [131]. VE deficiency is associated with prediabetes in apparently healthy individuals [132]. In clinical trials VE supplementation has been shown to protect residual beta-cell function in insulin-dependent diabetes mellitus [133, 134].

8.8 Vitamin K (VK)

Vitamin K (VK) is the generic name for a group of fat-soluble vitamins with a common structure that includes phylloquinones (VK1) and menaquinones (VK2) [135] and can regulate glycemic status [136–138]. Menaquinone-4 is the predominant form of VK2, which is present in large amounts in the pancreas [139]. Human studies indicate that VK-dependent protein osteocalcin [140], anti-inflammatory properties, and lipid-lowering effects may mediate the beneficial function of vitamin K in insulin sensitivity and glucose tolerance [140]. Glucose-stimulated insulin secretion is higher in pancreatic islet cells that have been treated with VK2 [139]. Vitamin K supplementation has been found to be associated with significant reductions in blood glucose, increased fasting serum insulin, reduced hemoglobin A1c and increased ß-cell function in diabetic animal studies [137].

9. Drugs and medications for diabetes

There are several classes of drugs and medications used to treat or manage diabetes [141, 142]. Here is a summary of the most common types:

Metformin is a biguanide medication that is often prescribed as the first-line treatment for type 2 diabetes [143]. It helps to lower blood glucose levels by reducing glucose production in the liver and improving insulin sensitivity [144].

Sulfonylureas are a class of medications that stimulate insulin secretion from the pancreas, helping to lower blood glucose levels. They close ATP-sensitive K-channels in the beta-cell plasma membrane. This closure causes depolarization of the cell membrane, which triggers the release of insulin into the bloodstream [145]. Examples of sulfonylureas include glyburide, glipizide, and glimepiride [146].

Dipeptidyl Peptidase-4 (DPP-4) inhibitors, also known as gliptins, are a class of medications that increase insulin secretion and decrease glucagon secretion, helping to lower blood glucose levels. They work by blocking the action of the enzyme DPP-4, which is responsible for breaking down incretin hormones such as GLP-1 (glucagon-like peptide 1) and GIP (glucose-dependent insulinotropic peptide) [147]. By blocking DPP-4, gliptins increase the levels of these hormones in the body. GLP-1 and GIP stimulate the secretion of insulin from the beta cells in the pancreas, reduce the production of glucose by the liver, and slow the emptying of food from the stomach. Examples of gliptins include sitagliptin, saxagliptin, and linagliptin [148].

GLP-1 receptor agonists are a class of medications that mimic the action of incretin hormones, stimulating insulin secretion, suppressing glucagon secretion, and slowing down gastric emptying. Examples include exenatide, liraglutide, and dulaglutide [149].

SGLT2 inhibitors are a class of medications that work by blocking the reabsorption of glucose in the kidneys, leading to increased urinary glucose excretion and lower blood glucose levels. Examples include canagliflozin, dapagliflozin, and empagliflozin.

Insulin is a hormone that is essential for the regulation of blood glucose levels. People with T1D and some with type 2 diabetes may require insulin therapy to manage their condition. Insulin can be administered through injections or an insulin pump [150].

It's important to note that diabetes treatment is highly individualized, and the choice of medication depends on a variety of factors, such as the type and severity of diabetes, other medical conditions, and the patient's preferences and lifestyle. A healthcare provider can help determine the most appropriate treatment plan for each individual.

10. Conclusion

Beta cell health is critical for the development and management of T1D, as these cells produce insulin, the hormone that regulates blood glucose levels. Current approaches to preserving and promoting beta cell health in T1D include optimizing blood glucose control, reducing inflammation, and minimizing exposure to environmental toxins that can damage beta cells. Scientists are also exploring the use of targeted immunotherapies to prevent or reverse the autoimmune destruction of beta cells.

Future directions in beta cell health and T1D treatment include the development of regenerative therapies aimed at restoring beta cell function. These therapies may involve the use of stem cells, gene editing techniques, or biomaterials that promote beta cell growth and survival.

In addition to these approaches, precision medicine and personalized care are also gaining momentum in the T1D field. This involves tailoring treatment plans to the specific needs and characteristics of individual patients, such as their genetic profile, immune system status, and environmental exposures.

Advances in technology are also transforming the landscape of T1D care and management. For example, continuous glucose monitoring systems and closed-loop insulin delivery systems are becoming increasingly sophisticated and may improve outcomes for individuals with T1D.

Overall, a multifaceted approach that includes a focus on preserving and promoting beta cell health, precision medicine, and technological advancements offers promise for improving outcomes and quality of life for individuals with T1D.

Author details

Sheila Owens-Collins
Lexington-Fayette County Health Department, American Academy of Pediatrics, Lexington, KY, United States

*Address all correspondence to: sowenscollins@gamil.com

References

[1] Mambiya M, Shang M, Wang Y, Li Q, Liu S, Yang L, et al. The play of genes and non-genetic factors on type 2 diabetes. Frontiers in Public Health. 2019;7:349

[2] Steck AK, Rewers MJ. Genetics of type 1 diabetes. Clinical Chemistry. 2011;**57**(2):176-185

[3] Bonnefond A, Unnikrishnan R, Doria A, Vaxillaire M, Kulkarni RN, Mohan V, et al. Monogenic diabetes. Nature Reviews. Disease Primers. 2023;**9**(1):12

[4] Low HC, Chilian WM, Ratnam W, Karupaiah T, Md Noh MF, Mansor F, et al. Changes in mitochondrial epigenome in type 2 diabetes mellitus. British Journal of Biomedical Science. 2023;**80**:10884

[5] Russell MA, Redick SD, Blodgett DM, Richardson SJ, Leete P, Krogvold L, et al. HLA class II antigen processing and presentation pathway components demonstrated by transcriptome and protein analyses of islet beta-cells from donors with type 1 diabetes. Diabetes. 2019;**68**(5):988-1001

[6] Bell GI, Horita S, Karam JH. A polymorphic locus near the human insulin gene is associated with insulin-dependent diabetes mellitus. Diabetes. 1984;**33**(2):176-183

[7] Balboa D, Saarimaki-Vire J, Borshagovski D, Survila M, Lindholm P, Galli E, et al. Insulin mutations impair beta-cell development in a patient-derived iPSC model of neonatal diabetes. eLife. 2018;7:e38519

[8] Modi H, Johnson JD. Folding mutations suppress early beta-cell proliferation. eLife. 2018;7:e43475

[9] Panova AV, Klementieva NV, Sycheva AV, Korobko EV, Sosnovtseva AO, Krasnova TS, et al. Aberrant splicing of INS impairs beta-cell differentiation and proliferation by ER stress in the isogenic iPSC model of neonatal diabetes. International Journal of Molecular Sciences. 2022;**23**(15):8824

[10] Sun J, Cui J, He Q, Chen Z, Arvan P, Liu M. Proinsulin misfolding and endoplasmic reticulum stress during the development and progression of diabetes. Molecular Aspects of Medicine. 2015;**42**:105-118

[11] Bottini N, Musumeci L, Alonso A, Rahmouni S, Nika K, Rostamkhani M, et al. A functional variant of lymphoid tyrosine phosphatase is associated with type I diabetes. Nature Genetics. 2004;**36**(4):337-338

[12] Steck AK, Baschal EE, Jasinski JM, Boehm BO, Bottini N, Concannon P, et al. rs2476601 T allele (R620W) defines high-risk PTPN22 type I diabetes-associated haplotypes with preliminary evidence for an additional protective haplotype. Genes and Immunity. 2009;**10**(Suppl. 1):S21-S26

[13] Zheng W, She JX. Genetic association between a lymphoid tyrosine phosphatase (PTPN22) and type 1 diabetes. Diabetes. 2005;**54**(3):906-908

[14] Begovich AB, Carlton VE, Honigberg LA, Schrodi SJ, Chokkalingam AP, Alexander HC, et al. A missense single-nucleotide polymorphism in a gene encoding a protein tyrosine phosphatase (PTPN22) is associated with rheumatoid arthritis. American Journal of Human Genetics. 2004;**75**(2):330-337

[15] Kyogoku C, Langefeld CD, Ortmann WA, Lee A, Selby S, Carlton VE, et al. Genetic association of the R620W polymorphism of protein tyrosine phosphatase PTPN22 with human SLE. American Journal of Human Genetics. 2004;**75**(3):504-507

[16] Smyth D, Cooper JD, Collins JE, Heward JM, Franklyn JA, Howson JM, et al. Replication of an association between the lymphoid tyrosine phosphatase locus (LYP/PTPN22) with type 1 diabetes, and evidence for its role as a general autoimmunity locus. Diabetes. 2004;**53**(11):3020-3023

[17] Howson JM, Dunger DB, Nutland S, Stevens H, Wicker LS, Todd JA. A type 1 diabetes subgroup with a female bias is characterised by failure in tolerance to thyroid peroxidase at an early age and a strong association with the cytotoxic T-lymphocyte-associated antigen-4 gene. Diabetologia. 2007;**50**(4):741-746

[18] Kavvoura FK, Ioannidis JP. CTLA-4 gene polymorphisms and susceptibility to type 1 diabetes mellitus: A HuGE review and meta-analysis. American Journal of Epidemiology. 2005;**162**(1):3-16

[19] Marron MP, Raffel LJ, Garchon HJ, Jacob CO, Serrano-Rios M, Martinez Larrad MT, et al. Insulin-dependent diabetes mellitus (IDDM) is associated with CTLA4 polymorphisms in multiple ethnic groups. Human Molecular Genetics. 1997;**6**(8):1275-1282

[20] Ueda H, Howson JM, Esposito L, Heward J, Snook H, Chamberlain G, et al. Association of the T-cell regulatory gene CTLA4 with susceptibility to autoimmune disease. Nature. 2003;**423**(6939):506-511

[21] Lowe CE, Cooper JD, Brusko T, Walker NM, Smyth DJ, Bailey R, et al. Large-scale genetic fine mapping and genotype-phenotype associations implicate polymorphism in the IL2RA region in type 1 diabetes. Nature Genetics. 2007;**39**(9):1074-1082

[22] Vella A, Cooper JD, Lowe CE, Walker N, Nutland S, Widmer B, et al. Localization of a type 1 diabetes locus in the IL2RA/CD25 region by use of tag single-nucleotide polymorphisms. American Journal of Human Genetics. 2005;**76**(5):773-779

[23] Guo D, Li M, Zhang Y, Yang P, Eckenrode S, Hopkins D, et al. A functional variant of SUMO4, a new I kappa B alpha modifier, is associated with type 1 diabetes. Nature Genetics. 2004;**36**(8):837-841

[24] Wang CY, She JX. SUMO4 and its role in type 1 diabetes pathogenesis. Diabetes/Metabolism Research and Reviews. 2008;**24**(2):93-102

[25] Lincez PJ, Shanina I, Horwitz MS. Reduced expression of the MDA5 gene IFIH1 prevents autoimmune diabetes. Diabetes. 2015;**64**(6):2184-2193

[26] Looney BM, Xia CQ, Concannon P, Ostrov DA, Clare-Salzler MJ. Effects of type 1 diabetes-associated IFIH1 polymorphisms on MDA5 function and expression. Current Diabetes Reports. 2015;**15**(11):96

[27] Burt RK, Oyama Y, Traynor A, Kenyon NS. Hematopoietic stem cell therapy for type 1 diabetes: Induction of tolerance and islet cell neogenesis. Autoimmunity Reviews. 2002;**1**(3):133-138

[28] Dadheech N, James Shapiro AM. Human induced pluripotent stem cells in the curative treatment of diabetes and potential impediments ahead. Advances in Experimental Medicine and Biology. 2019;**1144**:25-35

[29] Li L, Gu W, Zhu D. Novel therapy for type 1 diabetes: Autologous hematopoietic stem cell transplantation. Journal of Diabetes. 2012;**4**(4):332-337

[30] Nikoonezhad M, Lasemi MV, Alamdari S, Mohammadian M, Tabarraee M, Ghadyani M, et al. Treatment of insulin-dependent diabetes by hematopoietic stem cell transplantation. Transplant Immunology. 2022;**75**:101682

[31] Quinn LM, Wong FS, Narendran P. Environmental determinants of type 1 diabetes: From association to proving causality. Frontiers in Immunology. 2021;**12**:737964

[32] Drescher KM, von Herrath M, Tracy S. Enteroviruses, hygiene and type 1 diabetes: Toward a preventive vaccine. Reviews in Medical Virology. 2015;**25**(1):19-32

[33] Richardson SJ, Morgan NG. Enteroviral infections in the pathogenesis of type 1 diabetes: New insights for therapeutic intervention. Current Opinion in Pharmacology. 2018;**43**:11-19

[34] Tong M, Neusner A, Longato L, Lawton M, Wands JR, de la Monte SM. Nitrosamine exposure causes insulin resistance diseases: Relevance to type 2 diabetes mellitus, non-alcoholic steatohepatitis, and Alzheimer's disease. Journal of Alzheimer's Disease. 2009;**17**(4):827-844

[35] Alharbi HF, Algonaiman R, Alduwayghiri R, Aljutaily T, Algheshairy RM, Almutairi AS, et al. Exposure to bisphenol a substitutes, bisphenol S and bisphenol F, and its association with developing obesity and diabetes mellitus: A narrative review. International Journal of Environmental Research and Public Health. 2022;**19**(23):15918

[36] Li WZ, Stirling K, Yang JJ, Zhang L. Gut microbiota and diabetes: From correlation to causality and mechanism. World Journal of Diabetes. 2020;**11**(7):293-308

[37] Meijnikman AS, Gerdes VE, Nieuwdorp M, Herrema H. Evaluating causality of gut microbiota in obesity and diabetes in humans. Endocrine Reviews. 2018;**39**(2):133-153

[38] Ingrosso DMF, Primavera M, Samvelyan S, Tagi VM, Chiarelli F. Stress and diabetes mellitus: Pathogenetic mechanisms and clinical outcome. Hormone Research in Pædiatrics. 2023;**96**(1):34-43

[39] Berridge MJ. Vitamin D deficiency and diabetes. The Biochemical Journal. 2017;**474**(8):1321-1332

[40] Ali O. Genetics of type 2 diabetes. World Journal of Diabetes. 2013;**4**(4):114-123

[41] Lyssenko V, Groop L, Prasad RB. Genetics of type 2 diabetes: It matters from which parent we inherit the risk. The Review of Diabetic Studies. 2015;**12**(3-4):233-242

[42] Danaei G, Finucane MM, Lu Y, Singh GM, Cowan MJ, Paciorek CJ, et al. National, regional, and global trends in fasting plasma glucose and diabetes prevalence since 1980: Systematic analysis of health examination surveys and epidemiological studies with 370 country-years and 2.7 million participants. Lancet. 2011;**378**(9785):31-40

[43] Roep BO, Thomaidou S, van Tienhoven R, Zaldumbide A. Type 1 diabetes mellitus as a disease of the beta-cell (do not blame the immune system?). Nature Reviews. Endocrinology. 2021;**17**(3):150-161

[44] Li Y, Sun F, Yue TT, Wang FX, Yang CL, Luo JH, et al. Revisiting the antigen-presenting function of beta cells in T1D pathogenesis. Frontiers in Immunology. 2021;**12**:690783

[45] Saisho Y. Changing the concept of type 2 diabetes: Beta cell workload hypothesis revisited. Endocrine, Metabolic & Immune Disorders Drug Targets. 2019;**19**(2):121-127

[46] Gurgul-Convey E, Mehmeti I, Plotz T, Jorns A, Lenzen S. Sensitivity profile of the human EndoC-betaH1 beta cell line to proinflammatory cytokines. Diabetologia. 2016;**59**(10):2125-2133

[47] Wang J, Wang H. Oxidative stress in pancreatic beta cell regeneration. Oxidative Medicine and Cellular Longevity. 2017;**2017**:1930261

[48] Eguchi N, Vaziri ND, Dafoe DC, Ichii H. The role of oxidative stress in pancreatic beta cell dysfunction in diabetes. International Journal of Molecular Sciences. 2021;**22**(4):1509

[49] Leenders F, Groen N, de Graaf N, Engelse MA, Rabelink TJ, de Koning EJP, et al. Oxidative stress leads to beta-cell dysfunction through loss of beta-cell identity. Frontiers in Immunology. 2021;**12**:690379

[50] Swisa A, Glaser B, Dor Y. Metabolic stress and compromised identity of pancreatic beta cells. Frontiers in Genetics. 2017;**8**:21

[51] Zhu M, Liu X, Liu W, Lu Y, Cheng J, Chen Y. Beta cell aging and age-related diabetes. Aging (Albany NY). 2021;**13**(5):7691-7706

[52] Aguayo-Mazzucato C. Functional changes in beta cells during ageing and senescence. Diabetologia. 2020;**63**(10):2022-2029

[53] Eizirik DL, Colli ML, Ortis F. The role of inflammation in insulitis and beta-cell loss in type 1 diabetes. Nature Reviews. Endocrinology. 2009;**5**(4):219-226

[54] Nordmann TM, Dror E, Schulze F, Traub S, Berishvili E, Barbieux C, et al. The role of inflammation in beta-cell dedifferentiation. Scientific Reports. 2017;**7**(1):6285

[55] de Luca C, Olefsky JM. Inflammation and insulin resistance. FEBS Letters. 2008;**582**(1):97-105

[56] Inaishi J, Saisho Y. Beta-cell mass in obesity and type 2 diabetes, and its relation to pancreas fat: A mini-review. Nutrients. 2020;**12**(12):3846

[57] Patel P, Abate N. Body fat distribution and insulin resistance. Nutrients. 2013;**5**(6):2019-2027

[58] Bird SR, Hawley JA. Update on the effects of physical activity on insulin sensitivity in humans. BMJ Open Sport & Exercise Medicine. 2016;**2**(1):e000143

[59] Venturini M, Sallemi C, Marra P, Palmisano A, Agostini G, Lanza C, et al. Allo- and auto-percutaneous intra-portal pancreatic islet transplantation (PIPIT) for diabetes cure and prevention: The role of imaging and interventional radiology. Gland Surgery. 2018;7(2):117-131

[60] Bellin MD, Freeman ML, Gelrud A, Slivka A, Clavel A, Humar A, et al. Total pancreatectomy and islet autotransplantationinchronicpancreatitis: Recommendations from PancreasFest. Pancreatology. 2014;**14**(1):27-35

[61] Velazco-Cruz L, Goedegebuure MM, Millman JR. Advances toward engineering functionally mature human pluripotent stem cell-derived beta

cells. Frontiers in Bioengineering and Biotechnology. 2020;**8**:786

[62] Wan XX, Zhang DY, Khan MA, Zheng SY, Hu XM, Zhang Q, et al. Stem cell transplantation in the treatment of type 1 diabetes mellitus: From insulin replacement to beta-cell replacement. Frontiers in Endocrinology. 2022;**13**:859638

[63] Neumann M, Arnould T, Su BL. Encapsulation of stem-cell derived beta-cells: A promising approach for the treatment for type 1 diabetes mellitus. Journal of Colloid and Interface Science. 2023;**636**:90-102

[64] Shilleh AH, Russ HA. Cell replacement therapy for type 1 diabetes patients: Potential mechanisms leading to stem-cell-derived pancreatic beta-cell loss upon transplant. Cells. 2023;**12**(5):698

[65] Insel RA, Dunne JL, Atkinson MA, Chiang JL, Dabelea D, Gottlieb PA, et al. Staging presymptomatic type 1 diabetes: A scientific statement of JDRF, the Endocrine Society, and the American Diabetes Association. Diabetes Care. 2015;**38**(10):1964-1974

[66] Roder PV, Wu B, Liu Y, Han W. Pancreatic regulation of glucose homeostasis. Experimental & Molecular Medicine. 2016;**48**(3):e219

[67] Quesada I, Tudurí E, Ripoll C, Nadal Á. Physiology of the pancreatic α-cell and glucagon secretion: Role in glucose homeostasis and diabetes. The Journal of Endocrinology. 2008;**199**(1):5-19

[68] Mietlicki-Baase EG. Amylin-mediated control of glycemia, energy balance, and cognition. Physiology & Behavior. 2016;**162**:130-140

[69] Rorsman P, Huising MO. The somatostatin-secreting pancreatic delta-cell in health and disease. Nature Reviews. Endocrinology. 2018;**14**(7):404-414

[70] Kim W, Egan JM. The role of incretins in glucose homeostasis and diabetes treatment. Pharmacological Reviews. 2008;**60**(4):470-512

[71] Delamaire M, Maugendre D, Moreno M, Le Goff MC, Allannic H, Genetet B. Impaired leucocyte functions in diabetic patients. Diabetic Medicine. 1997;**14**(1):29-34

[72] Nitzan O, Elias M, Chazan B, Saliba W. Urinary tract infections in patients with type 2 diabetes mellitus: Review of prevalence, diagnosis, and management. Diabetes, Metabolic Syndrome and Obesity. 2015;**8**:129-136

[73] Blair Y, Wessells H, Pop-Busui R, Ang L, Sarma AV. Urologic complications in diabetes. Journal of Diabetes and its Complications. 2022;**36**(10):108288

[74] Wittig L, Carlson KV, Andrews JM, Crump RT, Baverstock RJ. Diabetic bladder dysfunction: A review. Urology. 2019;**123**:1-6

[75] Geerlings SE, Meiland R, van Lith EC, Brouwer EC, Gaastra W, Hoepelman AI. Adherence of type 1-fimbriated Escherichia coli to uroepithelial cells: More in diabetic women than in control subjects. Diabetes Care. 2002;**25**(8):1405-1409

[76] Mohanty S, Kamolvit W, Scheffschick A, Bjorklund A, Tovi J, Espinosa A, et al. Diabetes downregulates the antimicrobial peptide psoriasin and increases E. coli burden in the urinary bladder. Nature Communications. 2022;**13**(1):4983

[77] de Macedo GM, Nunes S, Barreto T. Skin disorders in diabetes mellitus: An

epidemiology and physiopathology review. Diabetology and Metabolic Syndrome. 2016;**8**(1):63

[78] Van Ende M, Wijnants S, Van Dijck P. Sugar sensing and signaling in Candida albicans and Candida glabrata. Frontiers in Microbiology. 2019;**10**:99

[79] Zhang S, Cai Y, Meng C, Ding X, Huang J, Luo X, et al. The role of the microbiome in diabetes mellitus. Diabetes Research and Clinical Practice. 2021;**172**:108645

[80] Rodrigues CF, Rodrigues ME, Henriques M. Candida sp. infections in patients with diabetes mellitus. Journal of Clinical Medicine. 2019;**8**(1):76

[81] De Santi F, Zoppini G, Locatelli F, Finocchio E, Cappa V, Dauriz M, et al. Type 2 diabetes is associated with an increased prevalence of respiratory symptoms as compared to the general population. BMC Pulmonary Medicine. 2017;**17**(1):101

[82] Pitocco D, Fuso L, Conte EG, Zaccardi F, Condoluci C, Scavone G, et al. The diabetic lung—A new target organ? The Review of Diabetic Studies. 2012;**9**(1):23-35

[83] Farmer MA, Linehan C, Marshall B. Does having diabetes mellitus increase the risk of developing active TB? Evidence-Based Practice. 2020;**23**(12):21-23

[84] Lee MR, Huang YP, Kuo YT, Luo CH, Shih YJ, Shu CC, et al. Diabetes mellitus and latent tuberculosis infection: A systematic review and metaanalysis. Clinical Infectious Diseases. 2017;**64**(6):719-727

[85] Liu Q, Yan W, Liu R, Bo E, Liu J, Liu M. The association between diabetes mellitus and the risk of latent tuberculosis infection: A systematic review and meta-analysis. Frontiers in Medicine. 2022;**9**:899821

[86] Yorke E, Atiase Y, Akpalu J, Sarfo-Kantanka O, Boima V, Dey ID. The bidirectional relationship between tuberculosis and diabetes. Tuberculosis Research and Treatment. 2017;**2017**:1702578

[87] Wei S, Li C, Wang Z, Chen Y. Nutritional strategies for intervention of diabetes and improvement of beta-cell function. Bioscience Reports. 2023;**43**(2):BSR20222151

[88] NIH. Vitamin A and Carotenoids Fact Sheet for Health Professionals. 2022 [updated June 15, 2022]. Available from: https://ods.od.nih.gov/factsheets/Vitamina-HealthProfessional/.

[89] Trasino SE, Benoit YD, Gudas LJ. Vitamin A deficiency causes hyperglycemia and loss of pancreatic beta-cell mass. The Journal of Biological Chemistry. 2015;**290**(3):1456-1473

[90] Iqbal S, Naseem I. Role of vitamin A in type 2 diabetes mellitus biology: Effects of intervention therapy in a deficient state. Nutrition. 2015;**31**(7-8):901-907

[91] Yang Y, Kim JW, Park HS, Lee EY, Yoon KH. Pancreatic stellate cells in the islets as a novel target to preserve the pancreatic beta-cell mass and function. Journal of Diabetes Investigation. 2020;**11**(2):268-280

[92] Zhou Y, Zhou J, Sun B, Xu W, Zhong M, Li Y, et al. Vitamin A deficiency causes islet dysfunction by inducing islet stellate cell activation via cellular retinol binding protein 1. International Journal of Biological Sciences. 2020;**16**(6):947-956

[93] Mascolo E, Verni F. Vitamin B6 and diabetes: Relationship and molecular

mechanisms. International Journal of Molecular Sciences. 2020;**21**(10):3669

[94] Marzio A, Merigliano C, Gatti M, Verni F. Sugar and chromosome stability: Clastogenic effects of sugars in vitamin B6-deficient cells. PLoS Genetics. 2014;**10**(3):e1004199

[95] Mascolo E, Amoroso N, Saggio I, Merigliano C, Verni F. Pyridoxine/pyridoxamine 5′-phosphate oxidase (Sgll/PNPO) is important for DNA integrity and glucose homeostasis maintenance in Drosophila. Journal of Cellular Physiology. 2020;**235**(1):504-512

[96] Qian B, Shen S, Zhang J, Jing P. Effects of vitamin B6 deficiency on the composition and functional potential of T cell populations. Journal of Immunology Research. 2017;**2017**:2197975

[97] Yahaya TO, Yusuf AB, Danjuma JK, Usman BM, Ishiaku YM. Mechanistic links between vitamin deficiencies and diabetes mellitus: A review. Egyptian Journal of Basic and Applied Sciences. 2021;**8**(1):189-202

[98] Lind MV, Lauritzen L, Kristensen M, Ross AB, Eriksen JN. Effect of folate supplementation on insulin sensitivity and type 2 diabetes: A meta-analysis of randomized controlled trials. The American Journal of Clinical Nutrition. 2019;**109**(1):29-42

[99] Zhao JV, Schooling CM, Zhao JX. The effects of folate supplementation on glucose metabolism and risk of type 2 diabetes: A systematic review and meta-analysis of randomized controlled trials. Annals of Epidemiology. 2018;**28**(4):249-57 e1

[100] Karampelias C, Rezanejad H, Rosko M, Duan L, Lu J, Pazzagli L, et al. Reinforcing one-carbon metabolism via folic acid/Folr1 promotes beta-cell differentiation. Nature Communications. 2021;**12**(1):3362

[101] NIH. Vitamin B12 Fact Sheet for Health Professionals. 2022. [updated December 22, 2022]. Available from: https://ods.od.nih.gov/factsheets/VitaminB12-HealthProfessional/.

[102] Green R, Allen LH, Bjorke-Monsen AL, Brito A, Gueant JL, Miller JW, et al. Vitamin B(12) deficiency. Nature Reviews. Disease Primers. 2017;**3**:17040

[103] Fotiou P, Raptis A, Apergis G, Dimitriadis G, Vergados I, Theodossiadis P. Vitamin status as a determinant of serum homocysteine concentration in type 2 diabetic retinopathy. Journal Diabetes Research. 2014;**2014**:807209

[104] Kim J, Ahn CW, Fang S, Lee HS, Park JS. Association between metformin dose and vitamin B12 deficiency in patients with type 2 diabetes. Medicine (Baltimore). 2019;**98**(46):e17918

[105] Mursleen MT, Riaz S. Implication of homocysteine in diabetes and impact of folate and vitamin B12 in diabetic population. Diabetes and Metabolic Syndrome: Clinical Research and Reviews. 2017;**11**(Suppl. 1):S141-S1S6

[106] NIH. Vitamin C Fact Sheet for Health Professionals. 2021. [updated March 26, 2021]. Available from: https://ods.od.nih.gov/factsheets/VitaminC-HealthProfessional/.

[107] Padayatty SJ, Levine M. Vitamin C: The known and the unknown and goldilocks. Oral Diseases. 2016;**22**(6):463-493

[108] Wilson R, Willis J, Gearry R, Skidmore P, Fleming E, Frampton C, et al. Inadequate vitamin C status in prediabetes and type 2 diabetes mellitus: Associations with glycaemic control, obesity, and smoking. Nutrients. 2017;**9**(9):997

[109] Ashor AW, Werner AD, Lara J, Willis ND, Mathers JC, Siervo M. Effects of vitamin C supplementation on glycaemic control: A systematic review and meta-analysis of randomised controlled trials. European Journal of Clinical Nutrition. 2017;**71**(12):1371-1380

[110] El-Aal AA, El-Ghffar EAA, Ghali AA, Zughbur MR, Sirdah MM. The effect of vitamin C and/or E supplementations on type 2 diabetic adult males under metformin treatment: A single-blinded randomized controlled clinical trial. Diabetes and Metabolic Syndrome: Clinical Research and Reviews. 2018;**12**(4):483-489

[111] Santosh HN, David CM, editors. Role of Ascorbic Acid in Diabetes Mellitus: A Comprehensive Review. 2017

[112] Dakhale GN, Chaudhari HV, Shrivastava M. Supplementation of vitamin C reduces blood glucose and improves glycosylated hemoglobin in type 2 diabetes mellitus: A randomized, double-blind study. Advances in Pharmacological Sciences. 2011;**2011**:195271

[113] NIH. Vitamin D Fact Sheet for Health Professionals. 2022 [updated August 12, 2022; cited 2023 February 12, 2023]. Available from: https://ods.od.nih.gov/factsheets/VitaminD-HealthProfessional/.

[114] Infante M, Ricordi C, Sanchez J, Clare-Salzler MJ, Padilla N, Fuenmayor V, et al. Influence of vitamin D on islet autoimmunity and beta-cell function in type 1 diabetes. Nutrients. 2019;**11**(9):2185

[115] Durmaz ZH, Demir AD, Ozkan T, Güçkan R, Tiryaki M. Does vitamin D deficiency lead to insulin resistance in obese individuals. Biomedical Research. 2017;**28**(17):7491-7497

[116] Cade C, Norman AW. Vitamin D3 improves impaired glucose tolerance and insulin secretion in the vitamin D-deficient rat in vivo. Endocrinology. 1986;**119**(1):84-90

[117] Mitri J, Pittas AG. Vitamin D and diabetes. Endocrinology and Metabolism Clinics of North America. 2014;**43**(1):205-232

[118] Norman AW, Frankel JB, Heldt AM, Grodsky GM. Vitamin D deficiency inhibits pancreatic secretion of insulin. Science. 1980;**209**(4458):823-825

[119] Tanaka Y, Seino Y, Ishida M, Yamaoka K, Yabuuchi H, Ishida H, et al. Effect of vitamin D3 on the pancreatic secretion of insulin and somatostatin. Acta Endocrinologica. 1984;**105**(4):528-533

[120] Szymczak-Pajor I, Sliwinska A. Analysis of association between vitamin D deficiency and insulin resistance. Nutrients. 2019;**11**(4):794

[121] Borissova AM, Tankova T, Kirilov G, Dakovska L, Kovacheva R. The effect of vitamin D3 on insulin secretion and peripheral insulin sensitivity in type 2 diabetic patients. International Journal of Clinical Practice. 2003;**57**(4):258-261

[122] Boucher BJ, Mannan N, Noonan K, Hales CN, Evans SJ. Glucose intolerance and impairment of insulin secretion in relation to vitamin D deficiency in East London Asians. Diabetologia. 1995;**38**(10):1239-1245

[123] Inomata S, Kadowaki S, Yamatani T, Fukase M, Fujita T. Effect of 1 alpha (OH)-vitamin D3 on insulin secretion in diabetes mellitus. Bone and Mineral. 1986;**1**(3):187-192

[124] Zhang D, Zhong X, Cheng C, Su Z, Xue Y, Liu Y, et al. Effect of vitamin D and/or calcium supplementation on

pancreatic beta-cell function in subjects with prediabetes: A randomized, controlled trial. Journal of Agricultural and Food Chemistry. 2023;**71**(1):347-357

[125] Al-Shoumer KA, Al-Essa TM. Is there a relationship between vitamin D with insulin resistance and diabetes mellitus? World Journal of Diabetes. 2015;**6**(8):1057-1064

[126] Nyomba BL, Auwerx J, Bormans V, Peeters TL, Pelemans W, Reynaert J, et al. Pancreatic secretion in man with subclinical vitamin D deficiency. Diabetologia. 1986;**29**(1):34-38

[127] Rasouli N, Brodsky IG, Chatterjee R, Kim SH, Pratley RE, Staten MA, et al. Effects of vitamin D supplementation on insulin sensitivity and secretion in prediabetes. The Journal of Clinical Endocrinology and Metabolism. 2022;**107**(1):230-240

[128] NIH. Vitamin E Fact Sheet for Health Professionals. 2021 [updated March 26, 2021]. Available from: https://ods.od.nih.gov/factsheets/VitaminE-HealthProfessional/.

[129] Parks E, Traber MG. Mechanisms of vitamin E regulation: Research over the past decade and focus on the future. Antioxidants & Redox Signaling. 2000;**2**(3):405-412

[130] Baburao Jain A, Anand JV. Vitamin E, its beneficial role in diabetes mellitus (DM) and its complications. Journal of Clinical and Diagnostic Research. 2012;**6**(10):1624-1628

[131] Sotoudeh G, Abshirini M, Bagheri F, Siassi F, Koohdani F, Aslany Z. Higher dietary total antioxidant capacity is inversely related to prediabetes: A case-control study. Nutrition. 2018;**46**:20-25

[132] Rodriguez-Ramirez G, Simental-Mendia LE, Carrera-Gracia MA, Quintanar-Escorza MA. Vitamin E deficiency and oxidative status are associated with prediabetes in apparently healthy subjects. Archives of Medical Research. 2017;**48**(3):257-262

[133] Crino A, Schiaffini R, Manfrini S, Mesturino C, Visalli N, Beretta Anguissola G, et al. A randomized trial of nicotinamide and vitamin E in children with recent onset type 1 diabetes (IMDIAB IX). European Journal of Endocrinology. 2004;**150**(5):719-724

[134] Pozzilli P, Visalli N, Cavallo MG, Signore A, Baroni MG, Buzzetti R, et al. Vitamin E and nicotinamide have similar effects in maintaining residual beta cell function in recent onset insulin-dependent diabetes (the IMDIAB IV study). European Journal of Endocrinology. 1997;**137**(3):234-239

[135] NIH. Vitamin K Fact Sheet for Health Professionals. 2021 [updated March 29, 2021]. Available from: https://ods.od.nih.gov/factsheets/VitaminK-HealthProfessional/.

[136] Ho HJ, Komai M, Shirakawa H. Beneficial effects of vitamin K status on glycemic regulation and diabetes mellitus: A mini-review. Nutrients. 2020;**12**(8):2485

[137] Karamzad N, Maleki V, Carson-Chahhoud K, Azizi S, Sahebkar A, Gargari BP. A systematic review on the mechanisms of vitamin K effects on the complications of diabetes and pre-diabetes. BioFactors. 2020;**46**(1):21-37

[138] Manna P, Kalita J. Beneficial role of vitamin K supplementation on insulin sensitivity, glucose metabolism, and the reduced risk of type 2 diabetes: A review. Nutrition. 2016;**32**(7-8):732-739

[139] Ho HJ, Shirakawa H, Hirahara K, Sone H, Kamiyama S,

Komai M. Menaquinone-4 amplified glucose-stimulated insulin secretion in isolated mouse pancreatic islets and INS-1 rat insulinoma cells. International Journal of Molecular Sciences. 2019;**20**(8):1995

[140] Al-Suhaimi EA, Al-Jafary MA. Endocrine roles of vitamin K-dependent-osteocalcin in the relation between bone metabolism and metabolic disorders. Reviews in Endocrine & Metabolic Disorders. 2020;**21**(1):117-125

[141] Feingold KR. Oral and injectable (non-insulin) pharmacological agents for the treatment of type 2 diabetes. In: Feingold KR, Anawalt B, Blackman MR, Boyce A, Chrousos G, Corpas E, et al., editors. South Dartmouth, MA: Endotext; 2000

[142] DeMarsilis A, Reddy N, Boutari C, Filippaios A, Sternthal E, Katsiki N, et al. Pharmacotherapy of type 2 diabetes: An update and future directions. Metabolism. 2022;**137**:155332

[143] Biguanides SG. A review of history, pharmacodynamics and therapy. Diabète & Métabolisme. 1983;**9**(2):148-163

[144] Pernicova I, Korbonits M. Metformin—mode of action and clinical implications for diabetes and cancer. Nature Reviews. Endocrinology. 2014;**10**(3):143-156

[145] Ashcroft FM. Mechanisms of the glycaemic effects of sulfonylureas. Hormone and Metabolic Research. 1996;**28**(9):456-463

[146] Costello RA, Nicolas S, Shivkumar A. Sulfonylureas. Treasure Island, FL: StatPearls; 2022

[147] Bohannon N. Overview of the gliptin class (dipeptidyl peptidase-4 inhibitors) in clinical practice. Postgraduate Medicine. 2009;**121**(1):40-45

[148] Kasina S, Baradhi KM. Dipeptidyl Peptidase IV (DPP IV) Inhibitors. Treasure Island, FL: StatPearls; 2022

[149] Drucker DJ. Mechanisms of action and therapeutic application of glucagon-like peptide-1. Cell Metabolism. 2018;**27**(4):740-756

[150] Donnor T, Sarkar S. Insulin-pharmacology, therapeutic regimens and principles of intensive insulin therapy. In: Feingold KR, Anawalt B, Blackman MR, Boyce A, Chrousos G, Corpas E, et al., editors. South Dartmouth, MA: Endotext; 2000

Chapter 3

Stem Cell-Derived Pancreatic Beta Cells for the Study and Treatment of Diabetes

Jessie M. Barra and Holger A. Russ

Abstract

Patients suffering from Type 1 Diabetes rely on the exogenous supply of insulin. Cell replacement therapy employing cadaveric islets cells has demonstrated a proof of principle for a practical cure, rendering patients insulin independent for prolonged periods of time. However, challenges remain before this innovative therapy can be widely accessed by diabetic patients. Availability of cadaveric donor islets is limited, necessitating the generation of an abundant source of insulin-producing pancreatic beta cells. Immunological rejection of the allogeneic transplant and recurring autoreactivity contribute to eventual graft failure in all transplant recipients. In the current chapter, we summarize past and current efforts to generate functional beta cells from pluripotent stem cells and highlight current knowledge on graft immune interactions. We further discuss remaining challenges of current cell replacement efforts and highlight potentially innovative approaches to aid current strategies.

Keywords: pancreas, insulin-producing beta cell, autoimmune diabetes, autoreactive T cell, human pluripotent stem cell, direct differentiation, genome engineering

1. Introduction

1.1 The beta cell and type 1 diabetes

Global regulation of the essential metabolite, glucose, relies on the proper action of small clusters of endocrine cells within the pancreas known as the islet of Langerhans. Mostly comprised of alpha, beta, delta, gamma, and epsilon cells, these endocrine populations release a tightly synchronized set of hormones to maintain systemic glycemic control [1, 2]. Most importantly beta cells respond to increased glucose levels by secreting the hormone insulin, which triggers glucose uptake by peripheral tissues such as the muscle and adipose tissue [3]. To counter regulate the beta cell's suppressive impacts on blood glucose levels, alpha cells produce the hormone glucagon, which acts on the liver to increase glucose production and raise glucose concentrations [4]. While less studied, literature suggests that somatostatin-producing delta cells, pancreatic polypeptide-producing gamma cells, and ghrelin-producing epsilon cells, also have counterregulatory impacts on hormone secretion from the beta and alpha cells and help provide satiety signals to modulate gastric processes [5–7]. When this delicate balance of hormonal signals is dysregulated, particularly when the release of insulin from beta cells is impaired, systemic glucose homeostasis is perturbed. The loss of functional insulin-producing beta cells results in chronic hyperglycemia, a key hallmark of the disease state known as diabetes mellitus.

Type 1 diabetes (T1D) is distinguished from other types of diabetes mellitus by the involvement of autoreactive immune cells in the destruction of the pancreatic beta cells. While T1D only accounts for 5–15% of total clinical cases of diabetes, incidence rates are steadily increasing worldwide [8]. Despite this concerning trend, the field currently lacks the ability to identify future T1D patients early in disease progression before glucose homeostasis is affected. Clinical diagnosis usually presents with severely elevated peripheral glucose measurements combined with positive titers for autoantibodies against known beta cell antigens such as insulin, glutamic acid decarboxylase-65 (GAD-65), islet antigen-2 (I-A2), or zinc transporter 8 (ZnT8) [9]. At the time of diagnosis, it is estimated that around 60–90% of the total beta cell mass has already been lost and circulating insulin levels are severely reduced [10]. Therefore, once identified, patients are dependent on exogenous insulin injections to maintain some glycemic control. Insulin can be administered through various technologies including insulin pens and closed-loop insulin systems or bi-hormonal pumps. However, the mental and financial burden of these therapies can be extremely costly to patients [11, 12]. While these innovative delivery approaches have significantly improved the precision of exogenous insulin, they fail to reach the exquisite responses provided by native pancreatic beta cells to the most subtle changes in systemic glucose levels. Therefore over time, severe secondary complications including nephropathy, neuropathy, cardiovascular disease, and retinopathy are likely [13]. In addition to potentially devastating long term affects experienced by patients with T1D, the constant risk of life-threatening hypoglycemic episodes, induced by administering too much insulin, is a considerable burden on quality of life. While not perfect, in most patients with T1D, exogenous insulin therapies are beneficial, but alternative approaches that better recapitulate native beta cell function are needed to significantly improve the health and welfare for patients with T1D.

1.2 Beta cell replacement

A promising treatment strategy to more faithfully restore euglycemia for patients with T1D is beta cell replacement therapy. The first true success was documented in 2000, when a group of investigators from the University of Alberta published their findings from a small clinical trial following seven patients undergoing cadaveric islet transplantation [14]. This protocol, which would come to be known as the Edmonton Protocol, succeeded in maintaining insulin independence within this small cohort of individuals for nearly 5 months after islet infusion into the portal vein of the liver and systemic nonsteroid immune suppression. Clinically, this procedure is nowadays most usually performed, although whole pancreas transplants are also performed. Islet transplantation is especially beneficial for individuals that are poorly controlled by or are unresponsive to traditional insulin therapies [15]. Islet transplantation awards significant benefits to patients with T1D, most notably protection from hypoglycemic events, complete or some independence from insulin injections, and overall improved quality of life, however donor availability and immunological challenges persist that limit widespread implementation of this therapy [16].

After utilization of the Edmonton Protocol at several prominent trial centers, the field of islet transplantation made significant strides as the various mechanisms that can contribute to islet graft rejection became better understood. One of the key hurdles thought to contribute to the ultimate failure of transplanted islets is immune-mediated rejection. This process begins within minutes after transplantation and involves islet-derived stress responses that quickly recruit and activate innate immune subsets [17, 18]. This acute reaction contributes significantly to early graft loss immediately after transplant but can also serve to activate the adaptive immune system.

During clinical islet transplantation into patients with T1D, unlike other organ replacement surgeries such liver transplants, the islet graft is at risk from two separate adaptive immune reactions: (1) genetic mismatch responses and (2) autoreactive responses against islet antigens. As transplanted islets are from a genetically distinct donor, the immune system within the recipient will mount a response against the islet graft in an allogeneic (or HLA-mismatch) context. This immunological response has been associated with failure of the islet graft, as detection of alloreactive T cells and humoral responses precedes the loss of graft function [19]. The use of both induction and maintenance immunosuppression regimens, such as anti-thymocyte globulin (ATG), are commonly utilized to combat alloreactive immune responses effectively [20]. However, the use of these immunosuppressive drugs can leave a patient at risk for opportunistic infections, often requiring the temporary cessation of immunosuppression to clear these infections, thus leaving the islet graft unprotected at times [19].

Additionally, unlike other transplantation contexts, patients with T1D have autoreactive immune responses to beta cell antigens. After the initial destruction of endogenous beta cells within the pancreas during the time of disease onset, a subset of autoreactive T and B cells is converted into long-lived memory cells [21]. These memory populations can persist within the patient for decades and can become reactivated after islet transplantation, when a new source of beta cell antigen is introduced. The recurrence of autoimmunity toward the islet graft has been correlated with reduced graft survival, even in the absence of detectable allogeneic responses [22, 23]. Commonly utilized immunosuppressive drug regimens are not effective in combating autoimmunity and may even induce proliferation of autoreactive memory T cells during islet transplantation [23, 24]. Therefore, to achieve successful beta cell replacement, immunosuppressive treatment strategies must not only suppress allograft rejection, but also prevent the reactivation of T-cell mediated autoimmunity.

As the study and understanding of these immune reactions has been expanded, recent clinical islet transplant studies can now achieve rates of insulin independence in upwards of 44% of patients 3 years post-transplantation with >90% of patients seeing complete protection from severe hypoglycemic events beyond 5 years post-transplantation [25]. Some of the key factors that have contributed to improved islet transplantation success include the processing of the islet preparations [26], islet culture and manipulation prior to transplantation [27], the acute treatment of the patient at the time of transplantation [20, 28], and the maintenance immunosuppression protocols utilized long term [16]. However, nearly all islet transplants eventually fail to maintain glycemic control, and this treatment strategy still suffers from a lack of abundant, high-quality cadaveric islets to treat all individuals with T1D using this approach. Therefore, alternative sources of functional insulin-producing beta cells and novel immunosuppressive approaches are needed to allow a widespread, effective implementation of cell replacement therapy.

1.3 Stem cells for regenerative medicine approaches

One of the most promising alternatives to limited cadaveric human islets for transplantation is to generate large pools of glucose responsive, insulin-producing cells from human stem cell sources (**Figure 1**). Stem cells are separated into five general categories according to the extent of their ability to differentiate into other cell types: totipotent, pluripotent, multipotent, oligopotent, and unipotent [29]. Totipotent cells can generate an entire organism from a single cell, encompassing all cell types found in the body, as well as extraembryonic tissues [30]. Pluripotent cells can differentiate into all cells of the three germ layers, endoderm, mesoderm, and ectoderm. Multipotent cells include all progenitor type cells that can differentiate into a limited number of cell types that are derived from that progenitor population, while oligopotent cells can generate cells of only one specific differentiation fate. Finally, unipotent cells are proliferating cells that can only produce more of the same cell type.

Human embryonic stem cells (hESC), a form of pluripotent stem cells (PSC) which are derived from the inner cell mass of *in vitro* fertilized embryos at the blastocyst stage, were first described in 1998 [31]. hESC can rapidly and indefinitely proliferate and differentiate into all cells of the human body, given appropriate signals are provided, thereby representing an attractive, quasi unlimited source of human cells that could be utilized for a multitude of regenerative medicine approaches [32]. To date, hundreds of hESC lines have been generated from donated, otherwise discarded blastocysts. Nevertheless, the use of human blastocysts to generate hESC has raised ethical concerns regarding their utilization in many countries.

The discovery of induced pluripotent stem cells (iPSC) generated from easily accessible patient tissues or cells such as skin, peripheral blood, and urine not

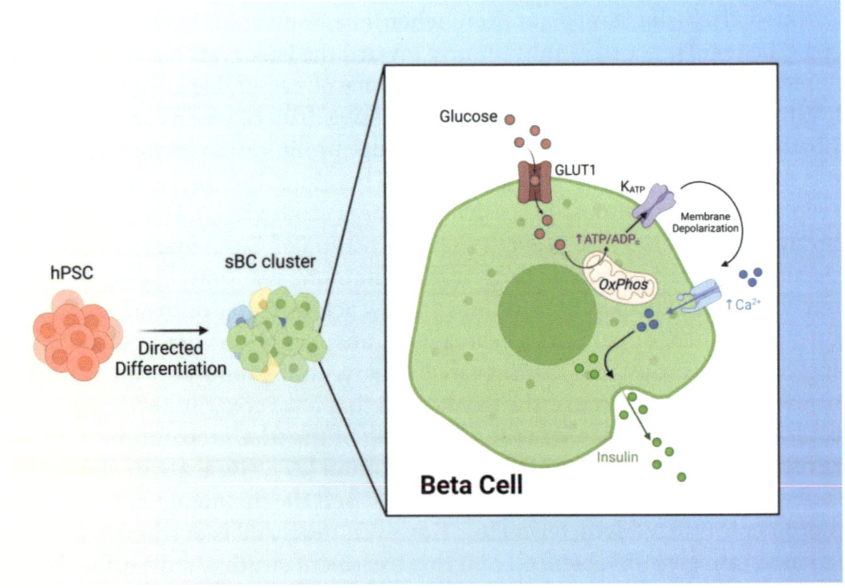

Figure 1.
Generating functional beta cells from human stem cells. Through a series of sequential stages, human pluripotent stem cells (hPSC) are directed differentiated into functional stem cell-derived beta-like cells (sBC). sBC can undergo glucose-stimulated insulin release. GLUT1; glucose transporter 1, OxPhos; oxidative phosphorylation, K_{ATP}; ATP-dependent potassium channels, Ca^{2+}; calcium. Figure generated using Biorender.

only lessened the ethical concerns surrounding human PSC but also allowed for increased genetic diversity by allowing the generation of patient specific iPSC [33–35]. The first successful attempt to reprogram somatic fibroblasts into iPSC utilized retroviral transduction of pluripotency genes Oct4, Sox2, Klf4, and c-myc into murine embryonic fibroblasts or tail-tip fibroblasts [36]. Similar approaches were utilized to reprogram human fibroblasts isolated from skin into human iPSC. Importantly, established iPSC display comparable features including differentiation potential to hESCs [33]. Collectively, either ESCs or iPSCs, represent a potentially unlimited supply of any desired cell type for cell replacement therapy, and therefore could address some of the challenges facing beta cell replacement strategies for the treatment of T1D.

2. Differentiation of stem cell-derived beta cells

To differentiate PSC into a pure population of beta cells, the delicate and time-specific process of pancreatic organogenesis must be closely replicated *in vitro*. The first attempts to generate insulin positive cells from PSC resulted in loosely defined pancreatic cell populations with some endocrine cells that were usually poly-hormonal and lacked the ability to secrete insulin in response to glucose [37]. Indeed, such differentiated cultures contained a fairly high number of non-endocrine populations such as pancreatic progenitors and exocrine cell types [38]. However, transplantation of these heterogenous pancreatic cells into immunodeficient mice promoted their differentiation into single hormone, insulin positive beta like cells after a few months. *In vivo* generated stem cell derived beta cells (sBC) are functional and secrete insulin in response to elevated glucose levels, providing the first proof of principle demonstration that PSC can differentiate into functional beta cells [39]. Furthermore, after chemical ablation of endogenous mouse beta cells using streptozotocin, human grafts containing functional sBC could maintain euglycemia, suggesting an important therapeutic potential to restore glycemic control [40, 41]. These early experiments provided evidence that directed differentiation of hPSC can produce functional beta cells, encouraging the research community to devise strategies to build upon earlier protocols with the goal of generating functional sBC *in vitro* [41–44]. Using these protocols, investigators have now better defined the key signals required to induce hPSCs commit to the appropriate differentiation trajectories in *vitro* that are essential to generate hormone positive endocrine cells.

2.1 From stem cell to endocrine progenitors

The first major fate decision needed to begin the process of directed differentiation is the induction of the endoderm lineage. WNT/β-catenin and the transforming growth factor beta (TGF-beta)/Nodal signaling pathways are the main drivers of endoderm differentiation [45]. Activin A, a member of a TGF-beta superfamily, is used *in vitro* to activate Nodal signaling. To initiate WNT signaling, CHIR-99021, an inhibitor of the suppressive kinase glycogen synthase kinase-3 beta (GSK3-beta), is nowadays commonly used. While initially researchers employed recombinant proteins to elicit certain signaling pathway modulation, small molecules provide more reproducible results and are far more cost effective. Overall, providing these signals in low serum media to hPSC induces the robust differentiation to endoderm, with upwards of 97% of cells upregulating key endoderm lineage markers [37].

Next, further differentiation to consecutively become primitive gut tube then posterior foregut are required. Fibroblast growth factor (FGF7) is frequently used to induce the differentiation of definitive endoderm to a primitive gut tube stage. In some instances, certain iPSC lines require additional low-grade BMP inhibition [46]. This differentiation step is followed by a combination of retinoic acid, Sant1 (a SHH signaling inhibitor), PMA (a PKC activator), and LDN (a BMP inhibitor) to further direct gut tube like cells toward both posterior foregut and then pancreatic progenitor stages [47]. Some beta cell differentiation protocols rely on the expansion of pancreatic progenitors before continuing with the induction of endocrine differentiation at this stage. For this purpose, pancreatic progenitors are cultured in the presence of epidermal growth factor (EGF) and FGF7 to establish a large pool of cells prior to endocrine induction [37, 39, 41–43, 47]. However, this amplification step is not necessarily required and extends the culture time required before reaching target beta cells considerably.

2.2 From pancreatic progenitor to mature beta cells

Current protocols induce endocrine cell differentiation from pancreatic progenitors by exposing cells to a cocktail of TGF-beta inhibition (ALK5i), thyroid hormone (T3), BMP inhibition (LDN), and NOTCH inhibition through gamma-secretase inhibition (XXi) [40, 41, 47–49]. This combination results in short lived expression of the master endocrine regulator gene NEUROG3 and activation of expression of downstream endocrine fate markers NKX2.1 and NEUROD1 [50]. Endocrine differentiation results in the generation of hormone expressing cells including sBCs of varying proportions based on the protocols utilized. While the resulting cell populations are heterogenous, the majority (greater than 90%) usually exhibit an early endocrine, non-proliferative phenotype, with sBCs representing commonly between 20 and 60% of all cells. Other hormone expressing cells are present at variable ratios and off target enterochromaffin cells, usually confined to the gut, are also detected at appreciable percentages (~10–15%), indicating that further protocol optimization might be required. Initially differentiated sBCs exhibit a rather immature phenotype [51] but further mature *in vitro* over the course of 1- several weeks to give rise to functional sBCs that exhibit several key features of primary human beta cells, including hormone and transcription factor expression, insulin granule formation and maturation, and elevated insulin secretion when exposed to elevated glucose concentrations [40, 41, 47]. Importantly, sBCs also display functionality shortly after transplantation, a distinct benefit when compared with the longer time period required for *in vivo* differentiation of transplanted pancreatic progenitors into sBC [52].

The WNT pathway as well as TGF-beta signaling have established roles in beta cell development and maturation, but these roles may be time dependent [48, 53, 54]. During beta cell development, inhibition of the TGF-beta receptor ALK5 is important for endocrine differentiation from pancreatic progenitors [48]. However during sBC maturation, inhibition or suppression of TGF-beta signaling results in sBCs with diminished insulin release [48]. Additionally, there is some evidence that canonical vs. non-canonical WNT signaling may influence the ability to induce optimal differentiation into functional sBCs. Different subpopulations of *in vitro* stem cell-derived definitive endoderm cells that rely on WNT signaling driven by either the canonical *WNT3A* ($CD275^+$ cells) or the non-canonical *WNT45A* and *WNT4* ($CD177^+$ cells) have altered capacity to specify toward the pancreatic lineage and therefore differed capacity to produce functional sBCs [55]. In this report, the non-canonical WNT signaling-driven DE population of CD177$^+$ cells had greater capacity to produce glucose responsive sBCs, suggesting that canonical WNT signaling may impair the commitment to the pancreatic cell fate [55]. Alternatively, other investigators have data indicating that the primary reason for the observed functional impairment in sBCs is a disruption in glycolytic processes [56]. This study suggested that GAPDH and PGK activity is altered in sBCs, resulting in decreased metabolite entry into the TCA cycle [56].

However, external signals also contribute to beta cell maturation. During mouse pancreatic organogenesis, newly derived endocrine cells including beta cells migrate and cluster together into the mini organ islet of Langerhans harboring the endocrine cells [57]. A similar phenomenon has also been observed during sBC maturation [58]. Insulin positive cells will group together within the heterogenous 3D cell cluster in a process referred to as "capping", resulting in improved function and increased expression of maturation markers MAFA and UCN3 [58]. Other laboratories have also confirmed that sorted insulin positive cells from later stages of a differentiation protocol will reaggregate together, leading to sBC with a more mature phenotype [38, 59]. These studies further highlight the importance of glucose metabolism and sBC architecture in functional maturation.

The timing and location of developmental signals are highly important aspects for pancreas organogenesis. Interactions with different nonendocrine cell populations such as endothelial cells or mesenchymal cells and the integration of both vascular and neuronal systems is also required to develop a proper functional pancreas, including the endocrine islets of Langerhans. Additionally, other non-beta cell endocrine populations within the islet have displayed the ability to more accurately fine tune the responsiveness of beta cells. Through release of molecules such as glucagon, glucagon-like peptide-1 (GLP-1), and acetylcholine, the alpha cell can enhance the secretion of insulin above that of glucose alone [60]. Somatostatin-producing delta cells also contribute to the regulation of beta cell function, as somatostatin is an inhibitor of both glucagon and insulin release [61]. A current potential barrier to efficient and consistent generation of hPSC-derived beta cells is that fully mature supporting cell types are largely absent form current direct differentiation approaches. Ongoing and future studies aim at incorporating multiple cell types to recapitulate *in vivo* developmental cell-cell interactions more closely under cell culture conditions. However, if the field lacks the ability to make fully mature beta cells, the real impact of the presence or absence of other cell types might be hard to assess.

Another potential confounding factor is that the current field's knowledge on pancreas organogenesis is largely derived from model systems, most notably frog, zebrafish, and mouse. Despite many key similarities between model systems and humans, a multitude of both structural and physiological differences between human and rodent islets have been described, leading to the recognition of vital differences in the complex dynamics of cell-cell interactions and the resultant insulin responsiveness of species specific beta cells [62]. This means that not all aspects of animal studies can or should be utilized as a reliable source of information that can be harnessed for the study of human physiology and associated model systems. To improve current protocols, further efforts should be focused on developing a comprehensive map of human beta cell development. This might be accomplished by combining an in-depth analysis of donated fetal pancreas tissues with an unbiased single cell or population analysis of both fetal and adult human samples. To accomplish this goal, PSC could represent a powerful tool to study human development in a controlled manner by providing access to otherwise inaccessible developmental timepoints. The hPSC differentiation model could provide key insights at later developmental stages since fetal pancreatic tissues past 20 weeks conception is not available. This should be done with caution, however, as many of the differentiation protocols used are aimed at driving the predominant phenotype of the beta cell, which could mask some of the delicate signaling balances that are surely present during organogenesis. More importantly, understanding the developmental signals and cues required to generate pancreatic cell types and potentially large pools of beta cells will accelerate our ability to treat human patients with type 1 diabetes.

3. Challenges to utilizing stem cell-derived beta cells for human disease

Once further optimized, generating large numbers of genetically identical and functional human beta cells in the laboratory will allow researchers to address the challenges facing cell replacement therapies in a reproducible manner. This possibility is improved by the rapidly evolving field of genome engineering that now allows for the establishment of novel human model systems that were previously largely restricted to animal models [63]. Since the first description of sBCs, researchers have proposed using these cells to address various aspects of beta cell replacement such as the ideal location to transplant, protecting grafts from ischemia during the peri-transplant period, shielding grafts from allogeneic and autoimmune recognition, and providing additional safety mechanisms to mitigate any off-target effects.

3.1 Generating immune-privileged beta cells

In addition to the optimization of directed differentiation of hPSCs into mature sBC, the field of T1D research has a vested interest in understanding the immune/beta cell interface [64]. This area of investigation includes trying to identify not only the early inflammatory events that lead to the activation of autoreactive immune cells, but also the consecutive interactions that ultimately cause specific beta cell destruction. Attaining such knowledge will not only help in protecting sBC grafts but also might provide insights as to how prevent autoimmunity directed against beta cells from developing in the first place. Precise and efficient genetic engineering usually harnesses targeted double strand DNA breaks to induce desired site-specific editing. Key advances in genome engineering have recently been described and technologies are centered on clustered regularly interspaced short palindromic repeats (CRISPR)/Cas9, Zinc finger, and Trans activator like nuclease families. Such genome engineering now enables efficient modification of hPSCs including the knockout (KO) of genes, insertion or knock-in (KI) of transgene cassettes, inducing point mutations, and changing the chromatin landscape locally to name a few of the most widely used applications [65].

Harnessing these approaches, various investigators have sought to generate hypo-immunogenic stem cell lines that could suppress the ability for cells to be detected by the immune system (**Figure 2**). Several groups have reported that HLA class I molecules are upregulated on the surface of beta cells in T1D organ donors and upon exposure to T1D mimicking, proinflammatory culture conditions [66–68]. HLA class I molecules are an integral piece of the process to activate cytotoxic $CD8^{+}$ T cells through their engagement with the CD8 T cell receptor. Therefore, genetically manipulating the expression of HLA class I molecules on beta cells could be an ideal target to abrogate both alloimmune and autoimmune cytotoxic responses toward beta cell grafts.

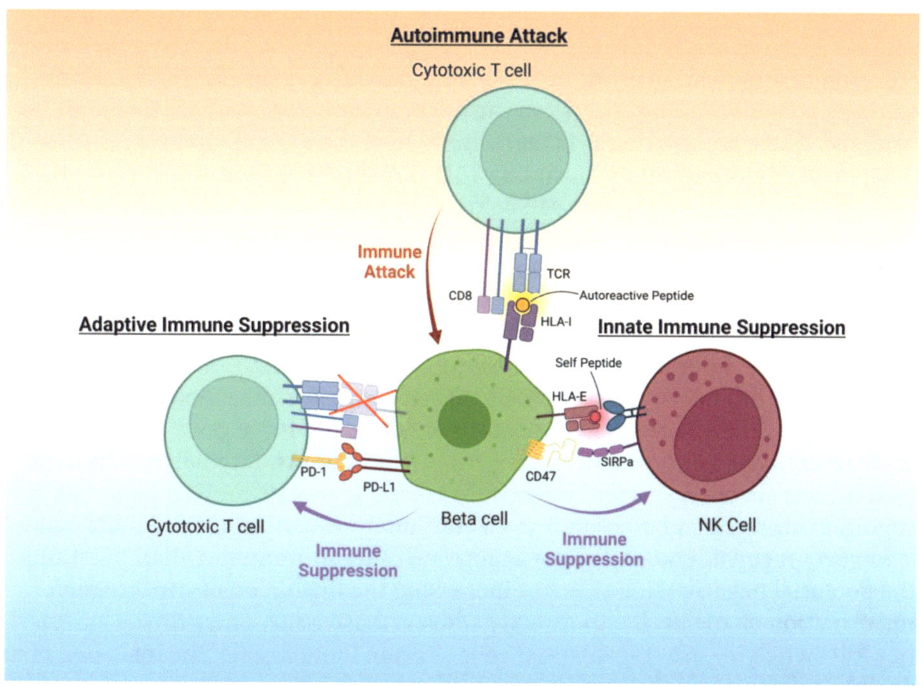

Figure 2.
Main mechanism of beta cell destruction and overview of approaches targeting immune suppression. An autoreactive T cell can recognize autoreactive peptides presented by the beta cell within HLA molecules. These HLA/peptide complexes interact with the autoreactive T cell receptor (TCR) and trigger the cytotoxic activity of the T cell. Through a deletion of all HLA-I molecules or overexpression of suppressive molecules such as programmed death-ligand 1 (PD-L1), this cytotoxic T cell action is suppressed. To also inhibit the innate immune-mediated destruction of the beta cell by natural killer (NK) cells, the specific expression of HLA-E and overexpression of inhibitory receptor CD47 have been utilized. Figure generated using Biorender.

To neutralize the negative impact of HLA class I expression on beta cells, various approaches have been assessed. hPSCs with a genetic knock out of the beta 2-microglobulin chain (B2M KO), shared among all HLA class I molecules has been utilized [67, 69]. B2M KO hPSC have been showed to exhibit reduced immune activation in an allogenic model [70, 71]. Generating B2M KO sBCs has demonstrated the ability to suppress autoreactive $CD8^{+}$ T cell activation in a fully matched human model system [67]. However, other studies have suggested an increase of natural killer (NK) cell-mediated cell death due to a lack of HLA-E/G expression required for normal immune surveillance by NK cells [72]. To overcome this challenge, lentiviral overexpression of HLA-E within B2M KO cells was able to suppress allogeneic T cell proliferation and activation without inducing NK cell activation [73]. As an alternative to maintaining HLA-E/G, another approach to reduce NK cell activity is the overexpression of CD47, which is a ubiquitously expressed immunomodulatory suppressive gene [74, 75]. CD47 is extremely effective at inhibiting NK cells killing of MHC-deficient iPSCs in immunocompetent mice [75].

Another popular method to suppress immune-mediated destruction of sBCs is the activation of the program cell death protein 1 and ligand (PD1/PD-L1) axis. In the normal adaptive immune response, PD1 is expressed on T and B cells, whereas its ligand PD-L1 is expressed on a variety of cells including antigen presenting cells [76]. The interaction between PD1 and PD-L1 results in a downregulation of the adaptive immune response [77]. To harness this suppressive pathway, overexpression of the ligand PD-L1 on hPSCs is utilized. PD-L1 overexpressing hPSCs displayed the ability to inhibit the activation of autoreactive immune cells *in vitro* [67, 78]. However, targeting a potent immunosuppressive pathway can have its downsides that must be considered. There are some reports that inhibition of the PD1/PD-L1 axis can cause the development of autoimmunity, including T1D [79–83]. Therefore, the safety of these approaches must be carefully assessed before they can be harnessed for clinical use.

Other investigators have sought to develop a strategy which addresses both adaptive and innate immune responses through a combination of genetic modifications to knockout the HLA molecule expression followed by knock-ins (KI) to express the immunomodulatory factors PD-L1, CD47 and HLA-G [70]. This study demonstrated that these modifications led to significant reduction in allogenic immune responses with respect to T cell, NK cell, and macrophage-mediated killing *in vitro* assays. Many of the above studies primarily investigated these genetic alterations in non-differentiated cell types using allogenic assay systems. It remains to be determined if these mechanisms will effectively protect stem cell-derived beta cells from recognition after transplantation into an autoimmune environment. Additionally, one important consideration for the utilization of hypo-immune cell populations is the potential negative side effect of increasing the likelihood of cancerogenic transformation of such cells. To directly address these issues and provide an emergency kill switch in case transplanted cells become tumorigenic, the inclusion of an inducible 'suicide gene' that can be triggered at will is likely essential for the safety of these approaches [84].

3.2 Modeling human beta cell/immune interactions

Once 'designer sBCs' have been generated, there is a critical need for ways to assess the *in vitro* and *in vivo* immune responses to this potential cell replacement therapy prior to clinical trials. Currently, the *in vitro* assessment of immune recognition and activation relies heavily on co-cultures of primary or PSC-derived human beta cells with allogenic human PBMCs. With the advent of iPSC technology this can now be done in a personalized, matched manner. Somatic cells from a T1D donor patient or control individual can be reprogrammed into iPSC followed by differentiation into sBCs. These sBC can then be co-cultured with immune cells isolated from the same respective donor. Control or T1D donor cells would be used to either exclude or include the autoreactive component of the immune response [70, 85]. However, such innovative co-culture approaches still have their limitations. Perhaps the most important of them being the fact that even within patients diagnosed with T1D, the frequency of autoreactive T cells in circulation is extremely low, preventing a proper assessment of the autoimmune recognition of sBCs with these models [86]. One method to circumvent reduced frequencies of autoreactive T cells is to genetically engineer autoreactive TCRs identified within the pancreas of T1D tissue donors into avatar T cells. This approach allows the generation of large numbers of T cells with known TCR reactivity for downstream co-culture experiments, improving access and reproducibility [87]. Notably, this approach allows for an HLA-matched assessment of autoreactive T cell destruction. Yet, even using these engineering approaches, *in vitro* cultures will only supply a partial, isolated understanding of the underlying mechanisms at play. Pre-clinical *in vivo* systems are also required to provide a more wholesome understanding of immune/beta cell interactions. While considerable strides have been made toward developing novel humanized animal models, further improvements are required, especially to accurately assess autoimmunity in these models.

Mice containing aspects of the human immune system, also known as 'humanized mice,' offer an amenable pre-clinical model of the human immune response and have been used for a variety of transplantation immunology studies [88–91]. There are number of humanized models available [92], but most advanced for transplant of hPSCs or their derivatives are those models which incorporate human hematopoietic stem cells (HSC) and thymic fragments into immune deficient mice to facilitate human T cell developmental in the animals. The bone-marrow-liver-thymus (BLT) model [93] and NeoThy model [94] are two popular models. Both contain de novo generated human HLA-restricted T cells, and a host of other adaptive and innate immune cell types useful for the assessment of transplantation success. Newer iterations include mouse models that have introduced mutations to various immunodeficient strains such as the NSG that remove the requirement for irradiation-based myeloablation before HSC transplant [95]. However, in the context of beta cell replacement therapies, these models only allow for the assessment of allogeneic graft rejection, therefore ignoring the key aspect of autoreactive destruction of a potential sBC graft. In addition, a common complication of humanized mice also present within these sophisticated models is the impaired trafficking of human T cells due to species differences of surface proteins.

To determine the true efficacy of a genetic alteration of sBCs, it is paramount to assess the autoimmune component. Potentially the closest animal system that currently exists to allow for the assessment of autoreactive immune cell recognition of a human sBC graft is a humanized model based on the non-obese diabetic (NOD) mouse. The NOD mouse is the most widely used pre-clinical T1D model, and unlike many other autoimmune disease models, the disease is spontaneous [95]. Importantly, disease risk is associated with numerous gene polymorphisms, many of them also found in T1D patients (*MHC-II*, *PTPN22*, *PTPN2*, *CTLA4*, *IL10*, *CTSH*, *CD226*, *IL2*, *RGS1*, *TAGAP*) [96]. To optimize the 'humanization' of this mouse model, all mouse MHC class I molecules were knocked out by genetic ablation of the conserved beta 2-microglobulin chain (NOD.β2m$^{-/-}$ mice) [97]. This deletion caused a resistance to the autoimmunity observed within normal NOD mice, highlighting again the importance of class I HLA molecules. Then, the most abundant T1D-associated class I susceptibility variant in humans HLA-A*02:01 was inserted (HHD). When introduced into normally disease-resistant NOD.β2m$^{-/-}$ mice, HHD transgene expression of HLA-A*02:01 restored the presence of pathogenic CD8^{+} T cells and T1D progression [98, 99]. These NOD-HLA-A2/HHD mice and their immunodeficient counterparts, NSG-HLA-A2/HHD mice, allow for a limited, but essential assessment of HLA-matched *in vivo* responses in an autoimmune environment. If using HLA-A*02:01 restricted hPSCs, this animal system allows for the assessment of autoimmune responses toward engineered sBCs [69]. However, presentation of mouse peptides during T cell development on human HLA-A2 molecules might still contribute a xenogeneic component to this model that should be considered when interpreting data. By continuously improving current humanized mouse models, the use of hPSC-derived beta cells along with humanized *in vitro* systems will allow for a true investigation of the human beta/immune cell interface in a matched autoimmune environment and open the door for a better understanding of possible clinical T1D interventions.

4. Summary

Since the initial development of directed differentiation protocols to generate human beta cells *in vitro,* the field has rapidly evolved. Early clinical trials involving transplantation of stem cell-derived pancreatic cells have shown some promising early results, reducing exogenous insulin requirements after transplantation [100]. Stem cell-derived beta cells can now be generated on a large scale, effectively removing the shortage of cadaveric islets previously preventing widespread utilization of cell therapy for T1D. First clinical trials using sBCs are currently in progress but peer reviewed reports as to the outcomes are still awaiting publication. While these are exciting times, seeing the implementation of stem cell technology for the treatment of diabetes in human patients, a few important challenges remain. Improvements to the directed differentiation process must be made to ensure the production of sufficient amounts of mature insulin-producing cells, as current protocols cannot achieve comparable glucose-responsive insulin release to cadaveric islets. Advances in the use of *in vitro* generated sBCs to model previously inaccessible aspects of human disease must be further explored, opening the potential for a better understanding of the human beta cell/immune interaction in the future. This will be critical knowledge to allow long term function of grafts. Harnessing the virtual limitless potential of genetic engineering to supply effective protection of sBC grafts from allogenic and recurring autoimmune rejection without systemic immunosuppression is an attractive prospect to allow for long-term survival and function of transplanted cells, and finally ensuring sufficient safety measures in case the engineered cells become aberrant are required for clinical trials. Addressing these remaining hurdles will allow for a well-tolerated beta cell replacement therapy for a large patient population affected by diabetes.

Author details

Jessie M. Barra[1,2] and Holger A. Russ[1,2]*

1 Diabetes Institute, University of Florida, Gainesville, FL, United States

2 Department of Pharmacology, University of Florida, Gainesville, FL, United States

*Address all correspondence to: holger.russ@ufl.edu

References

[1] Lammert E, Thorn P. The role of the islet niche on Beta cell structure and function. Journal of Molecular Biology. 2020;**432**:1407-1418

[2] Weitz J, Menegaz D, Caicedo A. Deciphering the complex communication networks that orchestrate pancreatic islet function. Diabetes. 2021;**70**:17-26

[3] Satoh T. Molecular mechanisms for the regulation of insulin-stimulated glucose uptake by small guanosine triphosphatases in skeletal muscle and adipocytes. International Journal of Molecular Sciences. 2014;**15**:18677-18692

[4] Ramnanan CJ, Edgerton DS, Kraft G, Cherrington AD. Physiologic action of glucagon on liver glucose metabolism. Diabetes, Obesity & Metabolism. 2011;**13**(Suppl 1):118-125

[5] Vijayan E, McCann SM. Suppression of feeding and drinking activity in rats following intraventricular injection of thyrotropin releasing hormone (TRH). Endocrinology. 1977;**100**:1727-1730

[6] Asakawa A et al. Characterization of the effects of pancreatic polypeptide in the regulation of energy balance. Gastroenterology. 2003;**124**:1325-1336

[7] cczaaaa SM, Page LC, Tong J. Ghrelin regulation of glucose metabolism. Journal of Neuroendocrinology. 2019;**31**:e12705

[8] Mobasseri M et al. Prevalence and incidence of type 1 diabetes in the world: A systematic review and meta-analysis. Health Promotion Perspective. 2020;**10**:98-115

[9] Kahanovitz L, Sluss PM, Russell SJ. Type 1 diabetes - A clinical perspective. Point Care. 2017;**16**:37-40

[10] Powers AC. Type 1 diabetes mellitus: Much progress, many opportunities. The Journal of Clinical Investigation. 15 Apr 2021;**131**(8):e142242

[11] Kesavadev J, Saboo B, Krishna MB, Krishnan G. Evolution of insulin delivery devices: From syringes, pens, and pumps to DIY artificial pancreas. Diabetes Therapy. 2020;**11**:1251-1269

[12] Blanchette JE, Toly VB, Wood JR. Financial stress in emerging adults with type 1 diabetes in the United States. Pediatric Diabetes. 2021;**22**:807-815

[13] Herman WH. The economic costs of diabetes: Is it time for a new treatment paradigm? Diabetes Care. 2013;**36**:775-776

[14] Shapiro AM et al. Islet transplantation in seven patients with type 1 diabetes mellitus using a glucocorticoid-free immunosuppressive regimen. The New England Journal of Medicine. 2000;**343**:230-238

[15] Choudhary P et al. Evidence-informed clinical practice recommendations for treatment of type 1 diabetes complicated by problematic hypoglycemia. Diabetes Care. 2015;**38**:1016-1029

[16] Rickels MR, Robertson RP. Pancreatic islet transplantation in humans: Recent Progress and future directions. Endocrine Reviews. 2019;**40**:631-668

[17] Nilsson B, Ekdahl KN, Korsgren O. Control of instant blood-mediated inflammatory reaction to improve islets of Langerhans engraftment. Current Opinion in Organ Transplantation. 2011;**16**:620-626

[18] Johansson H et al. Tissue factor produced by the endocrine cells of the islets of Langerhans is associated with a negative outcome of clinical islet transplantation. Diabetes. 2005;**54**:1755-1762

[19] Mohanakumar T et al. A significant role for histocompatibility in human islet transplantation. Transplantation. 2006;**82**:180-187

[20] Bellin MD et al. Potent induction immunotherapy promotes long-term insulin independence after islet transplantation in type 1 diabetes. American Journal of Transplantation. 2012;**12**:1576-1583

[21] Ehlers MR, Rigby MR. Targeting memory T cells in type 1 diabetes. Current Diabetes Reports. 2015;**15**:84

[22] Piemonti L et al. Alloantibody and autoantibody monitoring predicts islet transplantation outcome in human type 1 diabetes. Diabetes. 2013;**62**:1656-1664

[23] Monti P et al. Islet transplantation in patients with autoimmune diabetes induces homeostatic cytokines that expand autoreactive memory T cells. The Journal of Clinical Investigation. 2008;**118**:1806-1814

[24] Monti P, Piemonti L. Homeostatic T cell proliferation after islet transplantation. Clinical & Developmental Immunology. 2013;**2013**:217934

[25] Barton FB et al. Improvement in outcomes of clinical islet transplantation: 1999-2010. Diabetes Care. 2012;**35**:1436-1445

[26] Benomar K et al. Purity of islet preparations and 5-year metabolic outcome of allogenic islet transplantation. American Journal of Transplantation. 2018;**18**:945-951

[27] Froud T et al. Islet transplantation in type 1 diabetes mellitus using cultured islets and steroid-free immunosuppression: Miami experience. American Journal of Transplantation. 2005;**5**:2037-2046

[28] Koh A et al. Insulin-heparin infusions peritransplant substantially improve single-donor clinical islet transplant success. Transplantation. 2010;**89**:465-471

[29] Thanaskody K et al. MSCs vs. iPSCs: Potential in therapeutic applications. Front cell. Developmental Biology. 2022;**10**:1005926

[30] Posfai E et al. Evaluating totipotency using criteria of increasing stringency. Nature Cell Biology. 2021;**23**:49-60

[31] Thomson JA et al. Embryonic stem cell lines derived from human blastocysts. Science. 1998;**282**:1145-1147

[32] Martello G, Smith A. The nature of embryonic stem cells. Annual Review of Cell and Developmental Biology. 2014;**30**:647-675

[33] Takahashi K et al. Induction of pluripotent stem cells from adult human fibroblasts by defined factors. Cell. 2007;**131**:861-872

[34] Yu J et al. Induced pluripotent stem cell lines derived from human somatic cells. Science. 2007;**318**:1917-1920

[35] Nishikawa S-i, Goldstein RA, Nierras CR. The promise of human induced pluripotent stem cells for research and therapy. Nature Reviews Molecular Cell Biology. 2008;**9**:725-729

[36] Takahashi K, Yamanaka S. Induction of pluripotent stem cells from mouse

embryonic and adult fibroblast cultures by defined factors. Cell. 2006;**126**:663-676

[37] D'Amour KA et al. Efficient differentiation of human embryonic stem cells to definitive endoderm. Nature Biotechnology. 2005;**23**:1534-1541

[38] Veres A et al. Charting cellular identity during human in vitro β-cell differentiation. Nature. 2019;**569**:368-373

[39] Kroon E et al. Pancreatic endoderm derived from human embryonic stem cells generates glucose-responsive insulin-secreting cells in vivo. Nature Biotechnology. 2008;**26**:443-452

[40] Pagliuca FW et al. Generation of functional human pancreatic β cells in vitro. Cell. 2014;**159**:428-439

[41] Rezania A et al. Reversal of diabetes with insulin-producing cells derived in vitro from human pluripotent stem cells. Nature Biotechnology. 2014;**32**:1121-1133

[42] Schaffer AE et al. Nkx6. 1 controls a gene regulatory network required for establishing and maintaining pancreatic Beta cell identity. PLoS Genetics. 2013;**9**:e1003274

[43] Nostro MC et al. Efficient generation of NKX6-1+ pancreatic progenitors from multiple human pluripotent stem cell lines. Stem Cell Reports. 2015;**4**:591-604

[44] Millman JR et al. Generation of stem cell-derived β-cells from patients with type 1 diabetes. Nature Communications. 2016;**7**:1-9

[45] Nostro MC et al. Stage-specific signaling through TGFβ family members and WNT regulates patterning and pancreatic specification of human pluripotent stem cells. Development. 2011;**138**:861-871

[46] Nostro MC et al. Stage-specific signaling through TGFbeta family members and WNT regulates patterning and pancreatic specification of human pluripotent stem cells. Development. 2011;**138**:861-871

[47] Russ HA et al. Controlled induction of human pancreatic progenitors produces functional beta-like cells in vitro. The EMBO Journal. 2015;**34**:1759-1772

[48] Velazco-Cruz L et al. Acquisition of dynamic function in human stem cell-derived β cells. Stem Cell Reports. 2019;**12**:351-365

[49] Balboa D et al. Functional, metabolic and transcriptional maturation of human pancreatic islets derived from stem cells. Nature Biotechnology. 2022;**40**(7):1042-1055

[50] Sussel L et al. Mice lacking the homeodomain transcription factor Nkx2. 2 have diabetes due to arrested differentiation of pancreatic beta cells. Development. 1998;**125**:2213-2221

[51] Hrvatin S et al. Differentiated human stem cells resemble fetal, not adult, β cells. Proceedings of the National Academy of Sciences. 2014;**111**:3038-3043

[52] Vegas AJ et al. Long-term glycemic control using polymer-encapsulated human stem cell–derived beta cells in immune-competent mice. Nature Medicine. 2016;**22**:306-311

[53] Vethe H et al. The effect of Wnt pathway modulators on human iPSC-derived pancreatic beta cell maturation. Frontiers in Endocrinology. 2019;**10**:293

[54] Yoshihara E et al. Immune-evasive human islet-like organoids ameliorate diabetes. Nature. 2020;**586**:606-611

[55] Mahaddalkar PU et al. Generation of pancreatic beta cells from CD177(+) anterior definitive endoderm. Nature Biotechnology. 2020;**38**:1061-1072

[56] Davis JC et al. Glucose response by stem cell-derived β cells in vitro is inhibited by a bottleneck in glycolysis. Cell Reports. 2020;**31**:107623

[57] Adams MT, Blum B. Determinants and dynamics of pancreatic islet architecture. Islets. 2022;**14**:82-100

[58] Docherty FM et al. ENTPD3 marks mature stem cell derived beta cells formed by self-aggregation in vitro. Diabetes. Nov 2021;**70**(11):2554-2567. DOI: 10.2337/db20-0873

[59] Nair GG et al. Recapitulating endocrine cell clustering in culture promotes maturation of human stem-cell-derived β cells. Nature Cell Biology. 2019;**21**:263-274

[60] Moede T, Leibiger IB, Berggren PO. Alpha cell regulation of beta cell function. Diabetologia. 2020;**63**:2064-2075

[61] Hauge-Evans AC et al. Somatostatin secreted by islet delta-cells fulfills multiple roles as a paracrine regulator of islet function. Diabetes. 2009;**58**:403-411

[62] Rorsman P, Ashcroft FM. Pancreatic beta-cell electrical activity and insulin secretion: Of mice and men. Physiological Reviews. 2018;**98**:117-214

[63] Doudna JA, Charpentier E. The new frontier of genome engineering with CRISPR-Cas9. Science. 2014;**346**:1258096

[64] Pugliese A et al. The juvenile diabetes research foundation network for pancreatic organ donors with diabetes (nPOD) program: Goals, operational model and emerging findings. Pediatric Diabetes. 2014;**15**:1

[65] Hsu PD, Lander ES, Zhang F. Development and applications of CRISPR-Cas9 for genome engineering. Cell. 2014;**157**:1262-1278

[66] Bottazzo GF et al. In situ characterization of autoimmune phenomena and expression of HLA molecules in the pancreas in diabetic insulitis. New England Journal of Medicine. 1985;**313**:353-360

[67] Castro-Gutierrez R, Alkanani A, Mathews CE, Michels A, Russ HA. Protecting stem cell derived pancreatic Beta-like cells from diabetogenic T cell recognition. Frontiers in Endocrinology. 2021;**12**:707881

[68] Santini-Gonzalez J et al. Human stem cell derived beta-like cells engineered to present PD-L1 improve transplant survival in NOD mice carrying human HLA class I. Frontiers in Endocrinology (Lausanne). 2022;**13**:989815

[69] Wang D, Quan Y, Yan Q, Morales JE, Wetsel RA. Targeted disruption of the β 2-microglobulin gene minimizes the immunogenicity of human embryonic stem cells. Stem Cells Translational Medicine. 2015;**4**:1234-1245

[70] Han X et al. Generation of hypoimmunogenic human pluripotent stem cells. Proceedings of the National Academy of Sciences of the United States of America. 2019;**116**:10441-10446

[71] Wang B et al. Generation of hypoimmunogenic T cells from genetically engineered allogeneic human induced pluripotent stem cells. Nature Biomedical Engineering. 2021;**5**:429-440

[72] Gornalusse GG et al. HLA-E-expressing pluripotent stem cells escape allogeneic responses and lysis by NK cells. Nature Biotechnology. 2017;**35**:765-772

[73] Hoerster K et al. HLA class I knockout converts allogeneic primary NK cells into suitable effectors for "off-the-shelf" immunotherapy. Frontiers in Immunology. 2020;**11**:586168

[74] Advani R et al. CD47 blockade by Hu5F9-G4 and rituximab in non-Hodgkin's lymphoma. The New England Journal of Medicine. 2018;**379**:1711-1721

[75] Deuse T et al. Hypoimmunogenic derivatives of induced pluripotent stem cells evade immune rejection in fully immunocompetent allogeneic recipients. Nature Biotechnology. 2019;**37**:252-258

[76] Salmaninejad A et al. PD-1/PD-L1 pathway: Basic biology and role in cancer immunotherapy. Journal of Cellular Physiology. 2019;**234**:16824-16837

[77] Gianchecchi E, Delfino DV, Fierabracci A. Recent insights into the role of the PD-1/PD-L1 pathway in immunological tolerance and autoimmunity. Autoimmunity Reviews. 2013;**12**:1091-1100

[78] Ben Nasr M et al. PD-L1 genetic overexpression or pharmacological restoration in hematopoietic stem and progenitor cells reverses autoimmune diabetes. Science Translational Medicine. 2017;**9**:eaam7543

[79] Postow MA, Sidlow R, Hellmann MD. Immune-related adverse events associated with immune checkpoint blockade. New England Journal of Medicine. 2018;**378**:158-168

[80] Byun DJ, Wolchok JD, Rosenberg LM, Girotra M. Cancer immunotherapy—Immune checkpoint blockade and associated endocrinopathies. Nature Reviews Endocrinology. 2017;**13**:195-207

[81] Abdel-Wahab N, Shah M, Suarez-Almazor ME. Adverse events associated with immune checkpoint blockade in patients with cancer: A systematic review of case reports. PLoS One. 2016;**11**:e0160221

[82] Akturk HK et al. Immune checkpoint inhibitor-induced type 1 diabetes: A systematic review and meta-analysis. Diabetic Medicine. 2019;**36**:1075-1081

[83] Gauci ML et al. Autoimmune diabetes induced by PD-1 inhibitor-retrospective analysis and pathogenesis: A case report and literature review. Cancer Immunology, Immunotherapy. 2017;**66**:1399-1410

[84] Sułkowski M, Konieczny P, Chlebanowska P, Majka M. Introduction of exogenous HSV-TK suicide gene increases safety of keratinocyte-derived induced pluripotent stem cells by providing genetic "emergency exit" switch. International Journal of Molecular Sciences. 2018;**19**:197

[85] Leite NC, Pelayo GC, Melton DA. Genetic manipulation of stress pathways can protect stem-cell-derived islets from apoptosis in vitro. Stem Cell Reports. 2022;**17**:766-774

[86] Velthuis JH et al. Simultaneous detection of circulating autoreactive CD8+ T-cells specific for different islet cell-associated epitopes using combinatorial MHC multimers. Diabetes. 2010;**59**:1721-1730

[87] Landry LG et al. Proinsulin-reactive CD4 T cells in the islets of type 1 diabetes organ donors. Frontiers in Endocrinology (Lausanne). 2021;**12**:622647

[88] Shultz LD, Brehm MA, Garcia-Martinez JV, Greiner DL. Humanized mice for immune system investigation: Progress, promise and challenges. Nature Reviews. Immunology. 2012;**12**:786-798

[89] Rajesh D et al. Th1 and Th17 immunocompetence in humanized NOD/SCID/IL2rgammanull mice. Human Immunology. 2010;**71**:551-559

[90] Kalscheuer H et al. A model for personalized in vivo analysis of human immune responsiveness. Science Translational Medicine. 2012;**4**:125ra130

[91] Hermsen J, Brown ME. Humanized mouse models for evaluation of PSC immunogenicity. Current Protocols in Stem Cell Biology. 2020;**54**:e113

[92] Shultz LD et al. Humanized mouse models of immunological diseases and precision medicine. Mammalian Genome. 2019;**30**:123-142

[93] Lan P, Tonomura N, Shimizu A, Wang S, Yang YG. Reconstitution of a functional human immune system in immunodeficient mice through combined human fetal thymus/liver and CD34+ cell transplantation. Blood. 2006;**108**:487-492

[94] Brown ME et al. A humanized mouse model generated using surplus neonatal tissue. Stem Cell Reports. 2018;**10**:1175-1183

[95] McIntosh BE et al. Nonirradiated NOD,B6.SCID Il2rgamma−/− kit(W41/W41) (NBSGW) mice support multilineage engraftment of human hematopoietic cells. Stem Cell Reports. 2015;**4**:171-180

[96] Yu X, Huang Q, Petersen F. History and milestones of mouse models of autoimmune diseases. Current Pharmaceutical Design. 2015;**21**:2308-2319

[97] Serreze DV, Leiter EH, Christianson GJ, Greiner D, Roopenian DC. Major histocompatibility complex class I-deficient NOD-B2mnull mice are diabetes and insulitis resistant. Diabetes. 1994;**43**:505-509

[98] Takaki T et al. HLA-A*0201-restricted T cells from humanized NOD mice recognize autoantigens of potential clinical relevance to type 1 diabetes. Journal of Immunology. 2006;**176**:3257-3265

[99] Racine JJ et al. Improved murine MHC-deficient HLA transgenic NOD mouse models for type 1 diabetes therapy development. Diabetes. 2018;**67**:923-935

[100] De Klerk E, Hebrok M. Stem cell-based clinical trials for diabetes mellitus. Frontiers in Endocrinology. 2021;**12**:631463

Chapter 4

Physical Activity in Children and Adolescents with Type 1 Diabetes

Susan Giblin and Clodagh O'Gorman

Abstract

This chapter explores the multifaceted role of physical activity in type 1 diabetes management during childhood and adolescence. In addition to improved cardiovascular and metabolic health typically associated with physical activity, there are several diabetes-specific benefits of regular activity. For example, improved insulin sensitivity in insulin sensitivity that may be particularly important for children with type 1 diabetes approaching puberty when insulin resistance is known to increase, especially in females. Similarly, there are important diabetes-specific metabolic differences in response to physical activity that require consideration for blood glucose excursion management. Type, duration, and intensity of activity influence metabolic response in type 1 diabetes. For example, during aerobic activity, skeletal muscle glucose uptake increases to generate energy for muscle contraction, which suppresses hepatic gluconeogenesis and thus promotes a decrease in blood glucose levels and increased risk of hypoglycaemia. Intermittent, intense, or anaerobic activity can induce transient and often dramatic hyperglycaemia due to the release of the hormones epinephrine and glucagon. This rise in blood glucose can be followed by hypoglycaemia in the hours after activity. Within this chapter, the need for individualised and informed planning for safe participation in PA and exercise for children and adolescents is explored.

Keywords: physical activity, cardiovascular health, metabolic health, blood glucose excursion management, type 1 diabetes

1. Introduction

The global incidence of Type 1 Diabetes (T1D) in children is increasing; reports suggest that approximately 79,000 children develop T1D annually [1, 2]. T1D is a common, chronic, life-long illness with multifaceted considerations for the physical, psychological, and social implications associated with living with the condition [1]. For example, T1D is associated with increased risk of cardiovascular disease, neuropathy, and retinopathy. Similar to adult populations, the goal of T1D management in children is to promote health, maintain function, and to either prevent or delay adverse health outcomes, such as micro- and macrovascular changes, diabetic ketoacidosis, and renal impairment [3–5].

Insulin is the mainstay of management for T1D [2], non-pharmacological interventions that promote positive clinical, psychological, and social outcomes for chronic disease management have been recognised as an important addition to pharmacological management. Notably, physical activity (PA) is an important adjunct to insulin and dietary management for T1D [2–8]. 60 min of moderate to vigorous physical activity (MVPA) and limited sedentary time is recommended by The World Health Organization (WHO) for children to sustain health.

For healthy populations, there is a substantial amount of scientific evidence promoting the physical and psychological health benefits associated with living a physically active lifestyle [3–7]. PA has been associated with improvements in cardiovascular function, bone density, bone strength, musculoskeletal conditioning, blood pressure, insulin sensitivity, and blood lipid profiles [3–7]. PA also reduced the risk of comorbidities associated with sedentary lifestyles [3–7]. Consequently, promotion of PA with the aim of increasing PA engagement and reducing sedentary behaviour is the focus of the World Health Organization's global plan on PA (2018–2030) [3]. During childhood and adolescence, the experience of PA can shape future PA decisions and PA behaviours throughout later life [2–7]. Thus, it seems imperative and logical that early intervention promotes PA engagement in childhood and a pertinent component of diabetes management [3].

Individuals living with chronic illness or disability are at risk of not meeting PA recommendations due to limited safe, appropriate, and supportive PA initiatives to meet their additional needs. Unfortunately, children and adolescents with T1D often do not meet the WHO recommended PA targets. Despite the potential benefits of PA engagement [2–7], figures suggest that children with T1D are not meeting the recommended daily PA requirements to sustain health. PA engagement for children with T1D requires careful management of blood glucose excursions [2–7] and T1D populations face significant, disease-specific barriers to PA engagement.

Notably, most guidelines currently available for the support and promotion of PA in children and adolescents with type 1 diabetes are based on physiological knowledge and evidence derived from adult clinical studies. Further research is required to deepen our knowledge and understanding of PA in paediatric populations thus any exercise prescription or management plan should be individualised with prior experience and safety at the fore. Although medical complications are rare, medical clearance and guidance should be sought to support coaches and parents in determining any restrictions that may be relevant. For example, individuals who have proliferative retinopathy or nephropathy should avoid resistance-based exercise or anaerobic exercise that results in high arterial blood pressure.

For the purposes of this chapter, the terms PA, sport, and exercise are used interchangeable, however, PA typically refers to unstructured physical exertion. In contrast, sport and exercise typically refers to planned and structure activity that may include team or individual pursuits. PA, sport, and exercise can be further categorised depending on the nature of the activity and energy systems utilised. Throughout this chapter aerobic and anaerobic activities are discussed. Aerobic activity (e.g. walking, distance running) utilises oxygen and the cardiovascular system to provide energy to sustain the activity. Anaerobic activity (e.g. weight lifting, sprinting) is any activity that breaks down glucose for energy without using oxygen instead using lactic or alactic energy metabolism pathways.

2. Physiological principles of physical activity and diabetes

It is well established that PA and exercise are associated with numerous physiological benefits for sustaining health [2–7]. However, PA also presents a physiological and metabolic stressor that disrupts glycaemic balance and requires careful regulation to maintain homeostasis. To begin exploring the role of PA and exercise in type 1 diabetes management for children, it is important to consider the physiological differences (and similarities) in hormonal regulation at play. In healthy individuals, endogenous hormonal feedback and feedforward mechanisms underpin the maintenance of tightly controlled blood glucose levels. The hormone responsible for reducing blood glucose levels is insulin. There are several hormones that are responsible for increasing blood glucose levels in the body (e.g. glucagon, epinephrine, norepinephrine). During exercise, pancreatic beta cell production of insulin is suppressed, and pancreatic alpha cell production of glucagon is upregulated to manage the systemic supply of glucose in response to the physiological demands of the exercise or activity being undertaken. Muscle contraction also upregulates blood glucose transport into cells resulting in a counter regulatory reduction in circulating endogenous insulin levels in people without T1D. Individuals with T1D experience pathophysiological destruction of beta cells in the pancreas and thus lack the ability to tightly regulate blood glucose levels. Individuals with T1D rely on exogenous insulin, administered via pump or injection to ensure effective transport of glucose into cells. Exogenous insulin is not under strict endogenous feedback control mechanisms and thus peripheral insulin concentrations may rise during exercise due to increased mobilisation from the subcutaneous deposition and a reduction in insulin clearance.

In an individual without diabetes, glucose provision for exercise originates predominantly from the liver as a result of increased levels of glucagon and reduced circulating levels of insulin. However, during exercise in people with T1D, it is not possible to quickly change insulin levels and regulatory hormone responses can either increase or decrease as a consequence of activity [7, 9]. Such hormonal imbalances can be challenging to manage resulting in either hypo, hyper, or euglycaemia [7, 9].

The fundamental physiological principles that underpin glucose regulation during PA provide a foundation for understanding general blood glucose management principles for people with insulin dependent diabetes. However, a high level of individual variability exists in response to exercise, thus planning and managing exercise for people with T1D requires a highly personal approach. In addition to activity specific factors that influence glycaemic response, such as type, intensity, duration of activity (these factors will be discussed in more detail in later sections of this chapter), individual factors also need to be taken into consideration. For example, individual fitness level, initial blood glucose level, c-peptide secretion levels, and residual beta cell function influence glycaemic response during and after exercise or activity [9].

3. Type, duration, and intensity of physical activity and diabetes management

PA is commonly promoted as a method of improving glycaemic control, particularly in the context of the general population, however, there are important metabolic differences in response to PA between children with and without T1D [7]. Type, duration, and intensity of PA will influence metabolic response in T1D and careful planning is required to prevent instances of hyperglycaemia or hypoglycaemia [7, 9]. During aerobic activity, skeletal muscle glucose uptake increases to generate muscular activation. This increase in glucose uptake inhibits gluconeogenesis occurring in the liver and promotes reduction in blood glucose. Subsequently, the risk of hypoglycaemia is increased during aerobic activity. Given the risk of hypoglycaemia, it is pertinent to plan for aerobic activity of long duration or high intensity. Furthermore, exercise and PA is contraindicated for at least 24 hours following a severe hypoglycaemic event (i.e. hypoglycaemia resulting in cognitive impairment).

In addition to hypoglycaemia, hyperglycaemia can present a safety concern for undertaking certain types of PA or exercise. High intensity activity should not be undertaken if blood glucose levels are 14 mmol/L or above. Raised ketones present a safety concern prior to exercise and elevated ketones should be addressed prior to commencing exercise or PA, thus monitoring blood or urinary ketones is advised.

Adjustment to diabetes management regimens is typically required for any form of exercise lasting longer than 30 min [9]. For activity taking place during peak insulin activity time (i.e. soon after a meal), insulin dose reduction may be required. For example, reduction in short acting insulin is typically advised if given within a 2 hour period before exercise and supplementing with a snack if short acting insulin is given more than 2 hours before exercise. Further adjustments to basal insulin and bolus insulin may be required before and for several hours after exercise (e.g. throughout the night) for vigorous aerobic activity that has taken place in the evening. Basal insulin may also need to be reduced before bed depending on activity that has taken place earlier in the day. For insulin pump users, this may mean suspending pump activity temporarily. For those using multiple daily insulin injections, the site of injection may be an important consideration prior to activity. Large muscle groups that will be used during PA or exercise should be avoided as an injection site prior to activity [7, 9].

The requirement to plan for activity duration and intensity can be challenging, particularly for children where pitch-based activities and active play are spontaneous and involve repeated bouts of intense activity interspersed between rest or lower-level activity. Intermittent activity produces lesser reduction in blood glucose due to higher noradrenaline production [10, 11]. Similarly, due to changes in hormonal regulation, short bursts of anaerobic activities can lead to substantial increase in blood glucose. Elevated blood glucose in response to anaerobic activity is transient and can be followed by risk of hypoglycaemia for up to several hours after activity if not managed appropriately [12].

Additionally, the time of day that activity is undertaken can present challenges for children with T1D, nocturnal hypoglycaemia is common after PA during the day [13]. Risk of hypoglycaemia can remain elevated for up to 24 hours following activity due to increased insulin sensitivity associated with exercise.

Knowledge of physiological response to activity can also be a useful adjunct for regulating exercise mediated hypoglycaemia or hyperglycaemia. For example, risk of hypoglycaemia associated with prolonged aerobic activity can be reduced by the inclusion of high intensity bursts of activity (e.g. short sprints or strides) during or after activity. Conversely, hyperglycaemia associated with high intensity or anaerobic activity may be mitigated or reduced by the inclusion of a low-intensity aerobic 'cool down' activity. In the following section, commonly deployed strategies for managing and mitigating risk of hypoglycaemia and hyperglycaemia associated with exercise are discussed.

4. Strategies for PA promotion and T1D management

Insulin dose management around PA and exercise is an important component that requires specific education for parents, children, and adolescents. For individuals on multiple daily injection insulin regimens, reducing or missing basal insulin dose is not recommended to offset the increased insulin sensitivity and subsequent risk of hypoglycaemia in the hours following exercise cessation. Conversely, continuous insulin infusions (via pump technology) allow more flexibility for modifying insulin dosing, however, substantial reduction or suspension is not recommended and significant reductions in insulin dosing can lead to hyperglycaemic bouts. Furthermore, the optimal timing of insulin dose modification for exercise remains unclear [13].

Glucose monitoring is the mainstay of diabetes management. Traditionally self-monitoring of blood glucose is undertaken using finger prick assessment but increasingly technology is becoming more commonplace for tracking glucose levels in children and adolescents using real-time continuous glucose monitors (CGM). Information gathered from glucose monitoring allows refinement of future exercise strategies and can inform how different factors and behaviours influence blood glucose levels [13]. There are important differences to consider for self-monitoring of blood glucose levels via finger prick assessment and continuous glucose monitoring. For example, finger prick assessment measures glucose level directly from blood whereas continuous glucose monitoring devices track interstitial glucose levels. A lag can exist between blood and interstitial glucose, that may be pertinent to factor when reviewing glucose levels in response to activity. Consequently, blood glucose levels from continuous glucose monitoring devices may not be representative of decreasing or rapidly falling blood glucose during activity. However, continuous glucose monitors can provide valuable information about blood glucose levels before exercise and blood glucose excursions during and after activity allowing for ongoing refinement of individual-specific, tailored blood glucose management strategies. Unfortunately, some sports and activities may not be conducive to the use of continuous glucose monitoring, for example, contact sports. Different technologies also have different parameters to consider for use during water-based activities (e.g. depth of water, duration of immersion).

Blood glucose levels at the onset of exercise can be used to tailor glycaemic management strategies. 7–10 mmol/l is an acceptable starting range aerobic exercise for up to 60 min duration, however, expert opinion suggests that blood glucose target levels at the start of exercise should be individualised [13, 14]. It is important to note that in addition to starting blood glucose level, the typical rate of change in blood glucose should be taken into consideration to prevent hypoglycaemia. As discussed previously, the time since insulin dosing is another important consideration for activity and exercise management, particularly in relation to carbohydrate supplementation.

Carbohydrate supplementation is also commonly used to offset or reduce the risk of exercise-mediated hypoglycaemia. The risk of hypoglycaemia can often be managed through appropriate replacement of carbohydrate prior to, during, and after physical activity. Factors influencing the amount of carbohydrate intake required to prevent exercise-mediated hypoglycaemia include body mass, circulating insulin levels, and the type, intensity, and duration of exercise [7, 9, 13]. Clinical management with the multidisciplinary care team may include trial and error with both insulin dosing and carbohydrate supplementation to establish the optimal strategies for exercise. This can be particularly challenging for children and adolescents where exercise is not always planned and often spontaneous and intermittent in nature.

5. Psychosocial principles of physical activity and diabetes

Thus far, this chapter has explored the physiological and metabolic factors associated with exercise and T1D. In this section, the psychosocial concomitants of

PA participation are discussed. T1D management focus has progressed from solely being concerned with HbA1c optimisation to include holistic markers of both physiological and psychological health. There is a noted psychological burden associated with chronic disease management. The incidence of depression in children with T1D has been reported as three times higher than that of children without diabetes [15]. Exercise and PA may provide a mechanism for supporting psychosocial factors for children and adolescents that contribute to their quality of life and psychological well-being. Thus, it is important to consider the factors that support children and adolescents with T1D to participate in PA and the disease-specific barriers that they experience in leading a physically active life.

During childhood and adolescence, the experience of PA can shape future PA decisions and PA behaviours throughout later life [2–7]. The transition from childhood to adolescence is typically associated with a notable decrease in PA levels [2–7]. Therefore, PA experiences early in life can have life-long implications for health outcomes. The Childhood Determinants of Adult Health (CDAH) is a large-scale, longitudinal population-based study that reported the effect of behavioural, attitudinal, sociocultural, and physical factors on PA behaviours in healthy children [15]. For females, perceived competency and for males, physical fitness were found to be significant predictors of persisting with PA into adulthood. Early intervention to promote PA engagement seems imperative in childhood, particularly for children and adolescents with T1D who are at additional risk of low participation and increased sedentarism.

It is well established that in healthy populations, positive experiences in PA and exercise contribute to psychological wellbeing factors (e.g. enjoyment, confidence, and self-efficacy), however, it is important to note that diabetes-specific factors can influence the interaction between PA and psychological wellbeing for young people with T1D. Research has identified multiple barriers to PA engagement in individuals with T1D, fear of hypoglycaemia being reported most frequently. For individuals with T1D, exercise may mask symptoms of hypoglycaemia (e.g. tachycardia, diaphoresis, pallor). Whilst barriers to PA engagement are important to address through appropriate education and support, it is also important to understand the motivators and facilitators to PA engagement for individuals with T1D [16].

In keeping with Bandura's self-efficacy theory [17] and social cognitive theory [18], active peers and active role models are reported as being supportive of PA engagement for children with T1D. Self-efficacy theory purports that the belief that an individual can successfully perform an activity increases the likelihood of the individual to engage and persist in the activity. Patient compliance with exercise prescriptions is more likely to be successful if exercise self-efficacy is enhanced. Social cognitive theory is a behaviour theory of human motivation and action, that includes cognitive (e.g. self-efficacy) and environmental factors (e.g. social support) that interact with one another to shape human behaviour. Children who experience active families and friends are more likely to sustain an active lifestyle [19]. Self-efficacy and enjoyment are consistently acknowledged as important factors that help children with T1D to remain physically active. Conversely, low self-efficacy, anxiety, and lack of active peers contribute to low PA engagement. Healthcare professionals (HCP) working with children with T1D and their families have an important role to play in supporting and promoting PA. Evidence-based and individualised guidance from HCPs is important for managing the risk of hypoglycaemia and alleviating associated worries [20, 21]. As discussed earlier in this chapter, there are a number of tools and strategies available to aid in the planning and management of PA for children and adolescents (e.g. glucose monitoring, carbohydrate supplementation, insulin adjustment and technology). Each strategy and tool should be planned and managed in conjunction with a HCP team member providing diabetes care to children, adolescents, and their parents.

In addition to support provided in the clinical setting, international guidelines are available to support the promotion of opportunities for children and adolescents to safely participate in a variety sport, exercise, and PA settings. Unfortunately, to the best of our knowledge, there remains a lack of structured education initiatives aimed at transferring these guidelines in to practical application outside the healthcare setting, in schools, and sports clubs etc. Positively, general diabetes education programmes are available, for example, the International Society for Pediatric and Adolescent Diabetes (ISAPD) and International Diabetes Federation provide a structure education initiative for schools (KIDS programme) to promote education about diabetes and diabetes care [22]. Further specialised education initiatives that specifically address sport, physical activity, and exercise promotion for children and adolescents with T1D may help in increasing engagement and participation. The increased visibility of high level and elite sports people with diabetes is an important factor in encouraging and motivating young people with T1D to not only participate but to excel in sport and exercise endeavours.

6. Technology and physical activity

Technology for diabetes management is evolving rapidly. There are multiple modalities available to support day-to-day diabetes care, such as continuous glucose monitors, subcutaneous insulin infusion (insulin pumps), and closed loop technology. In addition to diabetes specific technologies, activity trackers and smartphone applications can provide additional information to support exercise management for individuals with diabetes. Technology allows accessible and transferable information that may be useful for paediatric diabetes management for example, a legal guardian, teacher, coach can access blood glucose information which may increase safety during and after PA or exercise. Some insulin pumps include algorithms that predict low glucose management that may be useful in mitigating or reducing the risk of exercise induced hypoglycaemia both during and after PA. Hybrid closed loop systems are now in use in dedicated paediatric diabetes clinic services and these automatically adjust insulin delivery in response to patterns of both hypoglycaemia and hyperglycaemia. It is important to develop and use individualised insulin patterns for exercise on these next generation insulin pumps [22, 23].

In addition to supporting the management of diabetes during PA and exercise, diabetes technology provides new insights into the impact of PA and exercise on acute and chronic markers of diabetes control. For example, while HbA1c was previously considered standard for monitoring optimal diabetes management, it may mask extremes in glucose variability. CGM provides a measure of 'time with range' or 'time in target', 'time above range' and 'time below range' for glucose targets. For accurate interpretation of CGM data, the percentage of time that the sensor is used is an important variable to monitor. Recent research, examining the acute impact of activity on CGM parameters of diabetes control, have provided further advocacy for the utility of moderate to vigorous activity to improve glycaemic control, with children achieving greater percentage time in range without significant time below or above range on days where greater activity levels are achieved [23].

Despite the rapidly evolving diabetes technologies, exercise and PA management remains one of the most significant challenges to automated systems. Technology-based management via continuous insulin infusion does appear to offer more flexibility and can reduce risk of post-exercise hyperglycaemia and delayed or nocturnal hypoglycaemia. However, highly individualised specific management and manual input are still required [13]. Unfortunately, exercise and sport may necessitate the removal/disconnection of pump infusions. Additionally, wearing a pump can present challenges to children and adolescents who may fear stigma or discrimination related to their diabetes.

Non-diabetes specific technology, such as wearable activity trackers that monitor heart rate, activity level, and intensity, can be used in conjunction with blood glucose monitoring, carbohydrate intake, and insulin dosing to better understand and support diabetes management in response to exercise and activity. Additionally, activity tracker technology provides objective, empirical information about parameters of activity (e.g. step count). Such objective information may reduce the challenges associated with 'self-report' approaches that are often used to assess and monitor physical activity behaviours in clinical settings. Minimum or 'target' step counts are often used as part of public health initiatives and physical activity promotion campaigns. For adults, 10,000 steps are commonly promoted as a daily target for sustaining physical health. For children and adolescents, there is some discrepancy between studies due to age categories, weight categories, and accelerometers/pedometers used, 11,500 steps per day have been identified as a target for both male and female children and adolescents [21, 24]. Activity trackers and smart phone applications can also provide important information about sedentary behaviour patterns, that are often under-reported or not assessed routinely despite the noted independent associations between sedentary time and physical health parameters.

The increasing availability of commercial technology has not only provided valuable clinically relevant information but has also changed the landscape of PA participation and promotion for the general healthy population. For example, smartphone applications and wearable activity tracker devices harness psychological and sociological concepts to support and motivate individuals to engage and persist in PA.

7. Conclusions

PA has potential to improve proactive and prophylactic management for children with T1D. In addition to physical and metabolic outcomes, PA has a positive impact on psychological aspects of living with chronic long-term conditions. PA and exercise management can present a challenge to caregivers, children, and adolescents as well as the multidisciplinary healthcare team, coaches, and teachers. Support and advice based on fundamental physiological principles of type, duration, and timing of PA and subsequent blood glucose response is needed to enable children and adolescents to participate safely in PA and exercise. PA requires careful, individualised planning, and monitoring to ensure appropriate insulin regimen or dietary modifications to reduce the risk of blood glucose excursions following activity. Prior to exercise, individuals with T1D should check their blood glucose levels prior to commencing activity, there should be ready access to glucose monitoring equipment and high glycaemic carbohydrate snacks available to treat hypoglycaemia.

Diabetes identification should be worn, and coaches, teachers, and teammates should be made aware of diabetes management requirements. Technology could play an important role in future research and PA promotion practices to aid in the transfer of PA guidelines to real-world changes in PA behaviours for children and adolescents with T1D.

Author details

Susan Giblin[1*] and Clodagh O'Gorman[2]

1 University Hospital Limerick, Limerick, Ireland

2 School of Medicine, University of Limerick, University Hospital Limerick, Limerick, Ireland

*Address all correspondence to: 18210252@studentmail.ul

References

[1] Atkinson MA, Eisenbarth GS, Michels AW. Type 1 diabetes. Lancet. 2014;**383**:69-82

[2] Ryden L, Grant PJ, Anker SD, et al. Task force on diabetes p-d, cardiovascular diseases of the European Society of C, European Association for the Study of D, ESC guidelines on diabetes, pre-diabetes, and cardiovascular diseases developed in collaboration with the EASD-summary. Diabetes & Vascular Disease Research. 2014;**11**:133-173

[3] Roche EF, McKenna A, Ryder K, Brennan A, O'Regan M, Hoey H. The incidence of childhood type 1 diabetes in Ireland and the National Childhood Diabetes Register. The Irish Medical Journal. 2014;**107**(9):278-281

[4] Chimen M, Kennedy A, Nirantharakumar K, Pang TT, Andrews R, Narendran P. What are the health benefits of physical activity in type 1 diabetes mellitus? A literature review. Diabetologia. 2012;**55**:542-551

[5] Michaud I, Henderson M, Legault L, Mathieu ME. Physical activity and sedentary behavior levels in children and adolescents with type 1 diabetes using insulin pump or injection therapy—The importance of parental activity profile. Journal of Diabetes and its Complications. 2017;**31**:381-386

[6] Trigona B, Aggoun Y, Maggio A, Martin XE, Marchand LM, Beghetti M, et al. Preclinical noninvasive markers of atherosclerosis in children and adolescents with type 1 diabetes are influenced by physical activity. The Journal of Pediatrics. 2010;**157**(4):533-539

[7] Herbst A, Bachran R, Kapellen T, Holl RW. Effects of regular physical activity on control of glycemia in pediatric patients with type 1 diabetes mellitus. Archives of Pediatrics & Adolescent Medicine. 2006;**160**(6):573

[8] Brazeau AS, Rabasa-Lhoret R, Strychar I, Mircescu H. Barriers to physical activity among patients with type 1 diabetes. Diabetes Care. 2008;**31**:2108-2109

[9] Taylor G, Smith K, Capper T, Scragg J, Bashir A, Flatt A, et al. Postexercise glycemic control in Type 1 diabetes is associated with residual β-cell function. Diabetes Care. 2020;**43**(10):2362-2370. DOI: 10.2337/dc20-0300

[10] Katz ML, Volkening LK, Anderson BJ, Laffel LM. Contemporary rates of severe hypoglycaemia in youth with type 1 diabetes: Variability by insulin regimen. Diabetic Medicine. 2012;**29**:926-932

[11] Patton SR, Dolan LM, Smith LB, Thomas IH, Powers SW. Pediatric parenting stress and its relation to depressive symptoms and fear of hypoglycemia in parents of young children with type 1 diabetes mellitus. Journal of Clinical Psychology in Medical Settings. 2011;**18**:345-352

[12] Patton SR, Dolan LM, Henry R, Powers SW. Fear of hypoglycemia in parents of young children with type 1 diabetes mellitus. 2008;**15**(3):252-259. DOI: 10.1007/s10880-008-9123-x

[13] Riddell MC, Gallen IW, Smart CE, Taplin CE, Adolfsson P, Lumb AN, et al. Exercise management in type 1 diabetes: A consensus statement. The Lancet: Diabetes and Endocrinology.

2017;**5**(5):377-390. DOI: 10.1016/S2213-8587(17)30014-1

[14] Maran A, Pavan P, Bonsembiante B, et al. Continuous glucose monitoring reveals delayed nocturnal hypoglycemia after intermittent high-intensity exercise in nontrained patients with type 1 diabetes. Diabetes Technology & Therapeutics. 2010;**12**:763-768

[15] Akbarizadeh M, Naderifar M, Ghaljaei F. Prevalence of depression and anxiety among children with type 1 and type 2 diabetes: A systematic review and meta-analysis. World Journal of Pediatrics. 2022;**18**:16-26

[16] Giblin S, Scully P, Dalton N, et al. Parent and child perceptions of physical activity with type 1 diabetes. BMJ Open Diabetes Research and Care. 2022;**10**:e002977. DOI: 10.1136/bmjdrc-2022-002977

[17] Braun V, Clarke V. Using thematic analysis in psychology. Qualitative Research in Psychology. 2006;**3**:77-101

[18] Voss C, Ogunleye AA, Sandercock GRH. Physical activity questionnaire for children and adolescents: English norms and cut- off points. Pediatrics International. 2013;**55**:498-507

[19] Gal JJ, Li Z, Willi SM, Riddell MC. Association between high levels of physical activity and improved glucose control on active days in youth with type 1 diabetes. Pediatric Diabetes. 2022 Nov;**23**(7):1057-1063. DOI: 10.1111/pedi.13391

[20] Quirk H, Blake H, Dee B, Glazebrook C. "Having diabetes shouldn't stop them": Healthcare professionals' perceptions of physical activity in children with Type 1 diabetes. BMC Pediatrics. 2015;**15**:68. DOI: 10.1186/s12887-015-0389-5

[21] Quirk H, Blake H, Tennyson R, Randell T, Glazebrook C. Physical activity interventions in children and young people with type 1 diabetes mellitus: A systematic review with meta-analysis. Diabetic Medicine. 2014;**31**(10):1163-1173

[22] DiMeglio LA, Acerini CL, Codner E, et al. ISPAD clinical practice consensus guidelines 2018: Glycemic control targets and glucose monitoring for children, adolescents, and young adults with diabetes. Pediatric Diabetes. 2018;**19**:105-114. DOI: 10.1111/pedi.12737

[23] Ng S, Wright N, Yardley D, Campbell F, Randell T, Trevelyan N, et al. Real world use of hybrid-closed loop in children and young people with type 1 diabetes mellitus-a National Health Service pilot initiative in England. Diabetic Medicine. 2023;**40**(2):e15015

[24] Li C, Chen X, Bi X. Wearable activity trackers for promoting physical activity: A systematic meta-analytic review. International Journal of Medical Informatics. 2021;**152**:104487. DOI: 10.1016/j.ijmedinf.2021.104487

Chapter 5

Impaired Physiological Regulation of ß Cells: Recent Findings from Type 2 Diabetic Patients

Shahzad Irfan, Humaira Muzaffar, Imran Mukhtar, Farhat Jabeen and Haseeb Anwar

Abstract

Recent studies have emphasized the multiple aspects of non-coding micro-RNAs in the regulation of pancreatic ß cells in type 2 diabetic patients. Thus, highlighting the significance of non-coding regions of the genome in regulating pancreatic endocrine cells. Functional dysregulation of pancreatic endocrine cells increases the incidence of metabolic disorders in otherwise healthy individuals. A precise understanding of the molecular biology of metabolic dysregulation is important from cellular and clinical perspectives. The current chapter will highlight the important recent findings from type 2 diabetic human patients and aims to enhance our current understanding of ß cell pathophysiology from a clinical perspective for the development of novel therapeutic approaches to control this global incidence.

Keywords: diabetes mellitus, ß cells, glucose, insulin, micro-RNAs

1. Introduction

Diabetes mellitus in principle is a pathophysiological condition in which the ability of cells to metabolize glucose is compromised primarily because of insulin deficiency or compromised insulin signaling (**Figure 1**). Diabetes mellitus is considered a global issue as it is estimated to impact around 700 million individuals worldwide by the year 2040 [1]. Hyperglycemia is the subsequent aftermath of diabetes mellitus where an abnormally high concentration of glucose persists at cellular and plasma levels. The persistent high concentration of glucose results in the activation of the polyol pathway and formation of advanced glycation end products (AGEs) along with an increase in AGE receptors (RAGE) [2, 3]. Activation of the polyol pathway enhances the mitochondrial production of reactive oxygen species (ROS) and increases the cytosolic concentration of ROS [4].

Type 2 diabetes mellitus (T2DM) is the most common metabolic disorder occurring globally [5–7]. Diets rich in carbohydrates and fats along with lack of exercise are highlighted as major risk factors for the increasing incidence of T2DM globally [8].

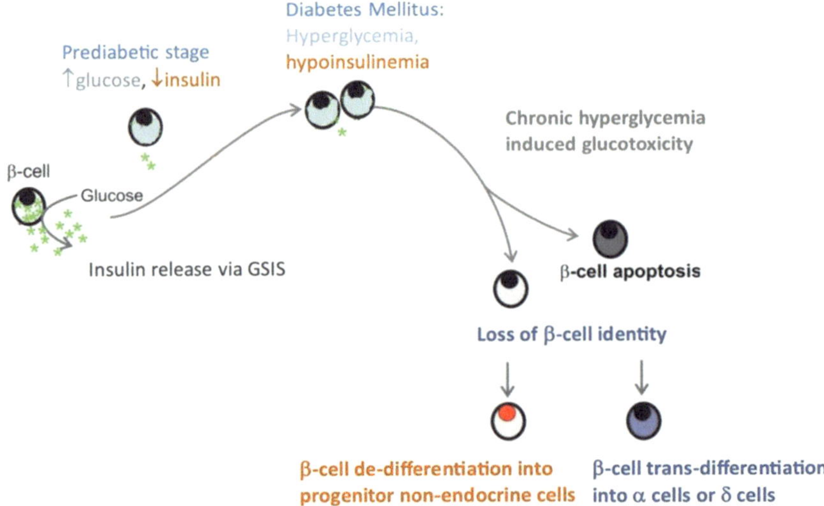

Figure 1.
Description of progressive ß cell decline during the course of diabetes mellitus and the subsequent loss of the cellular identity resulting in the trans- or de- differentiation of ß cells.

Obesity with visceral fat deposition and increased body mass index (BMI) have been shown to play a central role in the progression of type 2 diabetes complications [9].

2. Pathophysiology of type 2 diabetes mellitus (T2DM)

Regardless of the etiology, the T2DM progression is characterized either by a slow, progressive, and yet significant decline in the insulin secretion from ß cells or by a disruptive insulin signaling at the insulin-responsive tissues (muscles, fat, and liver) resulting in hyperinsulinemia which is clinically manifested as insulin resistance [10–13]. The inability of insulin to properly bind and activate insulin receptors (IRs) on the cell membrane of muscle cells, adipocytes, and hepatocytes has been attributed as the main reason for hyperinsulinemia and insulin resistance [14, 15].

As activation of cytosolic carbohydrate metabolism is induced by insulin and results in glucose phosphorylation and subsequent formation of glucose-6-phosphate inside the cells [16, 17]. Insulin signaling enhances the enzyme activities of hexokinases in muscle and fat tissue and glucokinase (GCK) activity in the ß cells and hepatocytes [18–21]. The activated cell surface insulin receptors stimulate cytosolic adaptor proteins called insulin receptor substrates (IRS1, IRS2) [22, 23]. Phosphorylation of IRS proteins activates the phosphoinositide2-kinase enzyme (PI3K) which further recruits ATP molecules to ultimately activate AKT protein (serine and threonine kinase) [24]. AKT activation helps the insulin to regulate multiple steps in glucose metabolism like a) increased cellular uptake of glucose via glucose transporter (GLUT4) in skeletal muscles [25], b) inhibition of glycogen synthase kinase 3 (GSK3) to stop glycogen metabolism [26–28], c) AKT induced activation of the mechanistic target of rapamycin (mTOR) resulting in the protein and lipid synthesis [29–34], d) transcriptional control of gene expression through AKT induced inhibition of forkhead family box O (FOXO) transcription factor proteins [35–37].

Insulin also greatly helps to reduce plasma fatty acid levels by reducing adipocyte lipolysis rate and enhancing the ability of hepatocytes to produce very low-density lipoprotein (VLDL) [38–43]. Insulin plays an important role in increasing skeleton muscle mass by enhancing the cellular intake of amino acids favoring protein synthesis and limiting the process of protein metabolism and urea formation [44–49].

These above-mentioned metabolic consequences of insulin signaling on macronutrients like carbohydrates, proteins, and lipids ensure the maintenance of nutritional homeostasis at the cellular level under diverse physiological conditions. Diabetes mellitus is fundamentally a loss of this metabolic homeostasis at the cellular level. The single and most important endogenous factor for the loss of metabolic homeostasis is dysfunctional insulin signaling. Once the insulin receptors are activated through binding insulin molecules, these receptors follow a deactivation phase by the process of internalization and thus can no longer bind with insulin, a phenomenon known as insulin receptor endocytosis [50]. Insulin receptor endocytosis is the primary physiological mechanism through which the duration and intensity of insulin signaling are controlled [51]. Hyperinsulinemia leads to insulin resistance by promoting insulin receptor endocytosis rate and diminishing the number of insulin receptors at the plasma membrane which could potentially bind with plasma insulin molecules [52]. Apart from receptor internalization, insulin receptor kinase activity is also shown to be compromised in T2DM patients [53]. Lack of or reduced insulin signaling fails to activate glucose transporter proteins (GLUT) which limit cellular uptake of glucose resulting in high plasma glucose levels. Normally high plasma glucose levels stimulate the release of insulin from ß cells, a phenomenon known as glucose-stimulated insulin secretion (GSIS) [54–56]. Surprisingly a compromised GSIS responsiveness of ß cells in terms of insulin secretion from T2DM donors has been reported [57–59]. Interestingly it has been reported that T2DM patients also have high plasma levels of glucagon along with insulin resistance, hyperinsulinemia, and hyperglycemia [60]. Highlighting that the insulin is unable to downregulate glucagon secretion from islet α cells [61]. T2DM patients have also been reported to have decreased levels of gut incretin hormone known as glucagon-like peptide-1 (GLP-1) which normally induces the postprandial release of insulin from ß cells and downregulates glucagon secretion from α cells [62, 63]. These findings implicate the gut as an important regulator of insulin secretion during feeding and fasting and might contribute to the incidence of T2DM [64, 65]. Current treatment options for T2DM include newly developed GLP-1 receptor agonists to curtail hyperglycemic episodes contributed by either the lack of insulin and/or increased glucagon secretion [66–68].

2.1 Transcriptional modulation of ß cells in T2DM

Altered gene expression profiles of important proteins involved in the secretory process of insulin and insulin signaling are the major consequence of T2DM, as *SUR1* and *TMEM37* (responsible for Ca^{2+} influx), *IR* (insulin receptor), and *GPD2, PCK1, ALDOB, FXYD2* (mitochondrial metabolism) have reduced expression in T2DM as compared to healthy controls [69–72]. Optimal mitochondrial activity is vital for normal insulin secretion because a normal ATP/ADP ratio promotes the K_{ATP} channels closure thus allowing Ca^{2+} influx and helping in the exocytosis of insulin granules from β cells. Mitochondrial ATP formation is also important for stimulating the conversion of proinsulin into insulin before insulin granule exocytosis.

Another important T2DM hallmark is the unusual loss of functional β cells, a process that starts during the prediabetic stage [73]. Almost half of the total population of functional β cells have been reported to be lost in T2DM patients [74–77]. Two distinct mechanisms have been credited for this remarkable loss in functional β cells during T2DM namely: 1) apoptosis and 2) dedifferentiation. Interestingly the process of β cell apoptosis is initiated due to the hyperactivity of β cells which results in hyperinsulinemia. Prolonged exposure of β cells to high levels of insulin activates enzymes like caspases which initiate apoptosis, and nitric oxide synthase (iNOS) triggering excessive nitric oxide (NO) production, along with the formation of hydrogen peroxide (H_2O_2) [78–80]. Hyperglycemia along with high plasma levels of free fatty acids, known as glucolipotoxicity, also induces β cell apoptosis in T2DM [81, 82]. The deleterious effect of increased lipid accumulation in the β cells includes hyperactivation of lipid signaling pathways causing loss of functional β cells. Recently cellular dedifferentiation (cells switching back to the undifferentiated/progenitor stage from the differentiated/functional stage) has emerged as an important factor influencing the functional β cell mass in T2DM [83–85]. β cell dedifferentiation also propagates the appearance of other islet cell types indicating that β cells might trans-differentiate into α and δ cell types. The removal of epigenetic control on the transcription of glucagon and somatostatin genes in β cells during trans-differentiation limits the unique identity of β cells as insulin secretory cells. This phenomenon of trans-differentiation might help to clarify the increased plasma levels of glucagon observed in T2DM patients. Specifically, glucagon and somatostatin-positive cells have inactivated (cytoplasmic) FOXO1 and NKX6.1 (β cell-specific transcription factors) proteins in the islets of T2DM donors hinting towards the possibility of trans-differentiation of β cells [84]. Transcriptionally upregulated pluripotent genes and downregulated β cell-specific genes have been observed to assist the trans- and dedifferentiation of functional β cells. Down-regulation of specific genes like *MafA, Nkx6.1,* and *FoxO1* which assign a unique identity to β cells, and upregulation of genes like *Ngn3, Oct4, Nanog,* and *L-Myc* which imparts cellular pluripotency has been observed in the islets of T2DM [86–90].

2.2 Epigenetic regulation of β cell-specific gene expression profile

Aberration in β cell functionality is mainly attributed to deregulated gene expression control through epigenetic mechanisms [91–96]. Specifically, chromatin modifications, DNA methylation, and post-translational modifications of histones are the classical epigenetic mechanisms through which gene expression profile of functional β cells are controlled. Alteration in the expression pattern of non-coding RNA sequences like microRNA (miRNA/miR) has also been shown to regulate the β cell function as well as their cellular identity [97–103]. Nonetheless, the recent findings that the majority of diabetes susceptibility loci are located in the non-coding regions of the human genome highlight the importance of epigenetic control in glucose homeostasis and β cell regulation [104, 105].

3. microRNA's regulation of β cell function

Pancreatic β cells are highly adaptative cellular entities in nature [106–109]. Adaptivity is specifically required to cater to different physiological states that demand different/opposing β cell responses. Like during feeding and fasting as well as during the high energy demands in exercise and pregnancy or during the intake of high carb high-fat diets. TF. Transcription upregulation of specific miRNAs during the embryonic developmental stage has been shown comprehensively to play an important part in the expression of a network of genes responsible for β cell development and function in mammals [110, 111]. However, the human data from diabetic adult patients suggests that the miRNAs might play a permissive role in the induction of β cell dysfunction [112]. On the other hand, certain miRNAs have been attributed to enhanced β cell mass/number in the pancreas, a phenomenon termed as β cell compensation [113–115]. These conflicting roles of miRNAs warrant a thorough understanding of the peculiar role of specific miRNAs under different physiological and pathological conditions. As miRNAs play a primary role in mRNA silencing and attenuate posttranscriptional regulation of gene expression in different physiological states of β cells. These fine-tuned variations are vital for glucose homeostasis. The loss of miRNA's ability to fine-tune the gene expression led to β cell decompensation which results in abnormal insulin secretion and ultimately the development of diabetes. β cell decompensation leads to two main pathophysiological events: 1) impaired (reduced) β cell function in terms of glucose sensitivity and insulin secretion (Reduced GSIS), and 2) reduced β cell number/mass due to dedifferentiation or apoptosis. Impaired β cell function involves molecular defects in insulin biosynthesis, glucose uptake, and exocytosis of insulin granules. The downregulation of β cell-specific genes and upregulation of the non β-cells specific genes induce β cell dedifferentiation. T2DM patients have been shown to present these β cell-specific conditions [116–118]. A set of specific molecular pathways are involved in the secretion of insulin during GSIS and failure of either one of these molecular pathways can result in the GSIS decline and consequently the incidence of diabetes.

3.1 Involvement of different miRNAs in insulin biosynthesis and signaling: Recent *ex-vivo* evidence from diabetic patients

The first comprehensive report on the possible involvement of miRNAs in regulating mammalian β-cells ability to synthesize and secrete insulin came in 2011. Down regulation of a subset of miRNA genes induced a decline in insulin gene promoter activity and subsequently resulted in reduced insulin content in isolated murine islets and cultured β cells [119]. These specific miRNAs were miRNA-24, miRNA-26, miRNA-148, and miRNA-181. Soon after another study employed pancreatic islets isolated from normal and glucose-intolerant human donors [120]. An islet-specific miRNA network involved in insulin secretion in human islets was suggested and consists of miRNA-375, miRNA-127-3p, and miRNA-184. Later, in 2015, miRNA-375 (an islet-specific miRNA) was shown to play a decisive role in regulating the phenotype of human β cells, and upregulation of miRNA-375 during *ex-vivo* expansion of human islet cells facilitates the generation of insulin-positive cells [121]. Experimental overexpression of miRNA-375 in dedifferentiated isolated human β cells also led to the restoration of β cell-specific gene expression profile. MafA (β cell-specific transcription factor regulating insulin gene) has been shown to be directly targeted by the miRNA-204 in isolated β cells from normal and diabetic human donors [122]. Briefly, Thioredoxin-interacting protein (TXNIP), a redox regulator intracellular protein, has been found to be upregulated in diabetes and induces β cell apoptosis. TXNIP upregulates the expression of miRNA-204 by inhibiting STAT3 activity. Overexpression of miRNA-204 decreases *MafA* expression in β cells thus lowering insulin production in diabetes. Followed by the report of over-expression of microRNA-124a in the islets isolated from T2DM patients [123]. microRNA-124a negatively regulates insulin by targeting mRNAs of key proteins involved in insulin gene transcription (NeuroD and FOXA2 and insulin signaling (Akt and Sirt1). miRNA-204 has been shown to downregulate β cell-specific GLP1 receptor expression in primary human islet cells by targeting GLP1R mRNA [124]. This finding might have therapeutic implications as the β cell-specific GLP1 receptor is responsible for the post-prandial insulin release in normal subjects and GLP-1 receptor agonists are used primarily as insulin secretagogues by T2DM patients [125, 126] same reference is present but from gut chapter]. Recent reports identify miRNA-181c-5p as a key factor in developing functional β cells through human-induced pluripotent stem cells (hiPSCs) by upregulating the expression of INS1, PDX1, NKX6.1, and MafA in [127]. More recently miRNA-7 has been attributed for its influence on insulin signaling via down-regulating the gene expression of insulin receptor substrate protein (*IRS1, IRS2*) during gestational diabetes in humans [128].

3.2 microRNAs influencing insulin secretion from β cells

As discussed earlier, the magnitude of glucose uptake during GSIS regulates the rate of insulin release from β cell. Currently multiple miRNAs have been identified in humans to affect the glucose uptake ability of the β cells thus indirectly reducing the rate of insulin release. Considerable decline in the expression of specific proteins responsible for K_{ATP} and Ca^{2+} channel regulation and insulin granular exocytosis has been observed in the islets from T2DM donors [129]. T2DM donor islets also present glucose induced rise in different miRNA levels namely miRNA-130a, miRNA-130b, and miRNA-152 [130]. These specific miRNAs impair post transcriptional mRNA processing of glucokinase and impacts glucose metabolism. Proper glucose metabolism via glucokinase action is essential for ATP synthesis inside β cells. As insulin exocytosis requires the closure of ATP sensitive K_{ATP} channels. On the contrary certain miRNAs have been shown to support β cells to preserve their unique cellular identity. Monocarboxylate transporter (MCT-1) is a mitochondrial protein which is specifically inhibited in β cells to allow them to only utilize glucose as a precursor for metabolism. These specific miRNAs include miRNA-29a, miRNA-29b, and miRNA-124 which have been shown to selectively target human MCT-1 gene [131]. miRNA-129a have also been shown to be glucose sensitive. As islets isolated from human donors when incubated in high glucose medium (hyperglycemic conditions) resulted in upregulation of miRNA-29 in β cells [132]. miRNA-29 also enhance the expression of *Onecut2* (a transcription factor). Onecut2 the granuphilin protein gene expression. Granuphilin protein blocks the exocytosis of secretory granules in endocrine cells. Apart from Onecut2-granuphilin mediated blocking of exocytosis, miRNA-29 also influences exocytotic protein genes like Syntaxin1A and SNAP25 (members of SNARE complex: a family of proteins involved in membrane fusion during exocytosis) [133].

4. microRNAs regulation by lipid accumulation in β cells

Another important and clinically relevant phenomenon of T2DM is the dysregulated lipid metabolism resulting in dyslipidemia (hypertriglyceridemia, reduced HDL and increased LDL particles) thus raising the incidence of obesity and cardiovascular disease (CVD) [134–136]. High plasma levels of triglycerides (hypertriglyceridemia) in T2DM patients result in substantial reduction of functional β cell numbers and significantly reduced insulin secretion [137]. Interestingly it has been demonstrated that microRNA sequences are altered during the β cell lipid metabolism by exposure to high lipids/fats [138–149]. Specific proteins responsible for lipid metabolism and cholesterol homeostasis in β cells have been found to be transcriptionally controlled by several microRNAs. miRNA-33 has been shown to regulate cholesterol homeostasis by controlling sterol regulatory binding protein (SREBP) gene expression [150]. miRNA-34a has been reported to induce β cell lipotoxicity *in*-vitro during the exposure of β cells to high concentrations of saturated fatty acids [151–154]. An increased cellular influx of fatty acids via di/triacylglycerol and/or esterified cholesterol pathways in β cell has been found to target an NAD^+-dependent deacetylase protein called sirtuin1 (SIRT1) [153]. SIRT1 is important for its specific role in the upregulation of insulin secretion in β cells in response to glucose stimulation [155]. Apart from insulin secretion, SIRT1 also transcriptionally controls multiple protein-encoding genes like tumor suppressor protein p53 and transcription factor proteins like nuclear factor κB (NF-κB) and FOXO family [156–158]. Apart from miRNA-34a, saturated fatty acids also impact the transcriptional profile of different miRNAs in the β cell contributing to lipotoxicity. Specific miRNAs regulated by saturated fatty acids include miRNA-146 [159], miRNA-182-5p [160], miRNA-297b-5p [161], and miRNA375 [162]. miRNA-146 targets mRNAs of certain genes responsible for the mediation of inflammatory pathways e.g. Toll-like receptors (TLRs), and NF-kB signaling [163]. INS-1 cells incubated with palmitate exhibit increased miRNA-182-5p expression, significantly decreased cell viability as well as increased palmitate-induced apoptosis. Interestingly, INS-1 cells when treated with specific inhibitors of miRNA-182-5p, exhibit a significant increase in cellular viability and reduction in palmitate-induced apoptosis [160]. Cultured human islets have increased levels of miRNA-146 in response to pro-inflammatory cytokines exposure decreased levels after high glucose

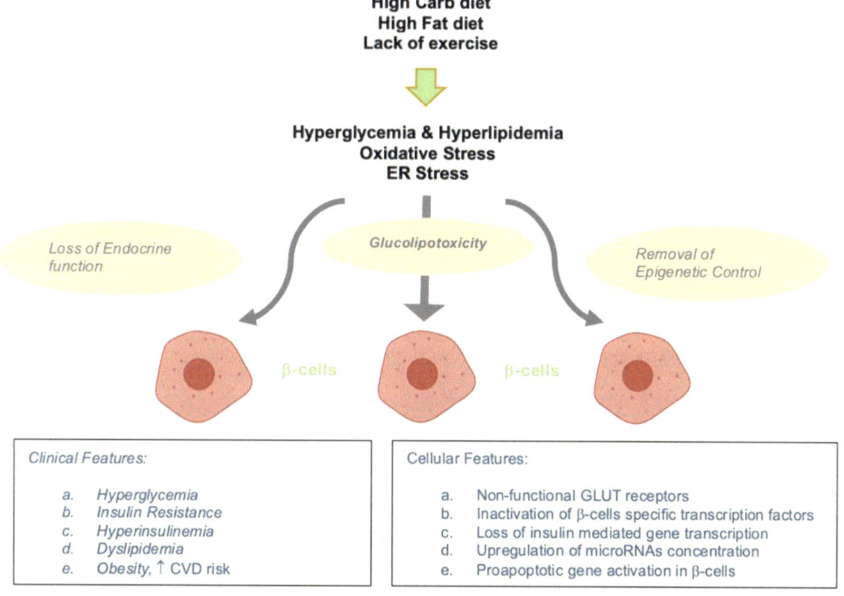

Figure 2.
Pathophysiology of type 2 diabetes mellitus.

exposure and interestingly stable levels after palmitate exposure [164]. Apart from the regulation of inflammatory response, miRNA-146 and miRNA 182-5p have also been involved in the reduction of cytoplasmic accumulation of lipids and cellular inflammatory response [165, 166]. β-cell lines exposed to stearic acid and palmitic acid exhibit reduced expression of miRNA-297b-5p whereas the increased miRNA-297b-5p levels help to minimize apoptosis induced by stearic acid but also result in reduced insulin concentration [161]. miRNA-375 has been shown to block high-fat diet-induced insulin resistance and obesity in mice by promoting hepatic expression of insulin-responsive genes [167] (**Figure 2**).

5. Conclusion

Epigenetic regulation of β cells supported by the growing number of high-impact studies on human patients have established the fact that microRNAs play a key role in defining β cell identity. Pathological dysregulation of β cells is in part caused by aberrant microRNA expression helping in the progression of T2DM. As the processes of microRNA biogenesis are characterized hinting towards the possibility of microRNAs as attractive therapeutic targets. Diabetes and β cell-specific microRNAs also represent distinct biomarkers for the early detection of diabetes. Because of the biochemical stability of microRNAs in the extracellular fluids like blood and plasma/serum, microRNAs also represent an excellent biomarker for diagnostics. Comprehensive studies involving large cohorts of diabetic patients to assess the predictive values of serum levels of different microRNAs in conjunction with the specific diabetic stage are required to delineate the pathological and beneficial role of specific microRNAs in diabetes mellitus.

Conflict of interest

"The authors declare no conflict of interest."

Abbreviation

T2DM	Type 2 diabetes mellitus
Micro-RNA	Micro- Ribo-Nucleic-Acid
ROS	Reactive Oxygen Species
AGE	Advanced Glycation End product
RAGE	Receptor of Advanced Glycation End Product
BMI	Body Mass Index
IRS	Insulin Receptor Substrate
GLUT	Glucose Transporter
GSIS	Glucose Stimulated Insulin Release
GCK	Glucokinase
FOXO	Forkhead Family Box O
GLP-1	Glucagon-like Peptide 1
GIP	Gastric Inhibitory Polypeptide
PDX1	Pancreatic Duodenal Homeobox 1
NAD	Nicotinamide Adenine Dinucleotide
VLDL	Very Low Density Lipoprotein

Author details

Shahzad Irfan[1*], Humaira Muzaffar[1], Imran Mukhtar[1], Farhat Jabeen[2] and Haseeb Anwar[1]

1 Department of Physiology, Govt College University, Faisalabad, Pakistan

2 Department of Zoology, Govt College University, Faisalabad, Pakistan

*Address all correspondence to: shahzadirfan@gcuf.edu.pk

References

[1] Maiese K. New insights for oxidative stress and diabetes mellitus. Oxidative Medicine and Cellular Longevity. 2015;**2015**:875961

[2] Brownlee M. The pathobiology of diabetic complications: A unifying mechanism (banting lecture). Diabetes. 2005;**54**(6):1615-1625

[3] Giacco F, Brownlee M. Oxidative stress and diabetic complications. Circulation Research. 2010;**107**(9):1058-1070

[4] Pi J, Zhang Q, Fu J, Woods CG, Hou Y, Corkey BE, et al. ROS signaling, oxidative stress and Nrf2 in pancreatic beta-cell function. Toxicology and Applied Pharmacology. 2010;**244**(1):77-83

[5] Hanley SC, Austin E, Assouline-Thomas B, Kapeluto J, Blaichman J, Moosavi M, et al. β-Cell mass dynamics and islet cell plasticity in human type 2 diabetes. Endocrinology. 2010;**151**(4):1462-1472

[6] Satin LS, Butler PC, Ha J, Sherman AS. Pulsatile insulin secretion, impaired glucose tolerance and type 2 diabetes. Molecular Aspects of Medicine. 2015;**42**:61-77

[7] Moin ASM, Dhawan S, Cory M, Butler PC, Rizza RA, Butler AE. Increased frequency of hormone negative and polyhormonal endocrine cells in lean individuals with type 2 diabetes. The Journal of Clinical Endocrinology and Metabolism. 2016;**101**(10):3628-3636

[8] Roden M, Shulman GI. The integrative biology of type 2 diabetes. Nature. 2019;**576**(7785):51-60

[9] Galicia-Garcia U, Benito-Vicente A, Jebari S, Larrea-Sebal A, Siddiqi H, Uribe KB, et al. Pathophysiology of type 2 diabetes mellitus. International Journal of Molecular Sciences. 2020;**21**(17):1-34

[10] Zick Y. Insulin resistance: A phosphorylation-based uncoupling of insulin signaling. Trends in Cell Biology. 2001;**11**(11):437-441

[11] Chatterjee S, Khunti K, Davies MJ. Type 2 diabetes. Lancet. 2017;**389**(10085):2239-2251

[12] Lewis GF, Carpentier A, Vranic M, Giacca A. Resistance to insulin's acute direct hepatic effect in suppressing steady-state glucose production in individuals with type 2 diabetes. Diabetes. 1999;**48**(3):570-576

[13] Weyer C, Bogardus C, Mott DM, Pratley RE. The natural history of insulin secretory dysfunction and insulin resistance in the pathogenesis of type 2 diabetes mellitus. The Journal of Clinical Investigation. 1999;**104**(6):787-794

[14] RA Haeusler TMDA. Biochemical and cellular properties of insulin receptor signaling. Nature Reviews. Molecular Cell Biology. 2018;**19**(1):31-44

[15] Boucher J, Kleinridders A, Ronald KC. Insulin receptor signaling in normal and insulin-resistant states. Cold Springer Harbor Perspective Biology. 2014;**6**(1):a009191

[16] Posner BI. Insulin signaling: The inside story. Canadian Journal of Diabetes. 2017;**41**(1):108

[17] Newsholme EA, Dimitriadis G. Integration of biochemical and physiologic effects of insulin on glucose metabolism. Experimental and Clinical Endocrinology & Diabetes. 2001;**109**(Suppl. 2):S122-S134

[18] Matschinsky FM, Wilson DF. The central role of glucokinase in glucose homeostasis: A perspective 50 years after demonstrating the presence of the enzyme in islets of Langerhans. Frontiers in Physiology. 2019;**10**:148

[19] Massa ML, Gagliardino JJ, Francini F. Liver glucokinase: An overview on the regulatory mechanisms of its activity. IUBMB Life. 2011;**6391**:1-6

[20] Iynedjian PB. Molecular physiology of mammalian glucokinase. Cellular and Molecular Life Sciences. 2009;**66**(1):27-42

[21] Matschinsky FM. Regulation of pancreatic-cell glucokinase: From basics to therapeutics. Diabetes. 2002;**51**(Suppl. 3):S394-S404

[22] Haeusler RA, McGraw TE, Accili D. Biochemical and cellular properties of insulin receptor signaling. Nature Reviews. Molecular Cell Biology. 2018;**19**(1):31-44

[23] White MF. IRS proteins and the common path to diabetes. American Journal of Physiology – Endocrinology Metabolism. 2002;**283**(3):46-43

[24] Kohn AD, Kovacina KS, Roth RA. Insulin stimulates the kinase activity of RAC-PK, a pleckstrin homology domain containing ser/thr kinase. The EMBO Journal. 1995;**14**(17):4288-4295

[25] Zorzano A, Palacín M, Gumà A. Mechanisms regulating GLUT4 glucose transporter expression and glucose transport in skeletal muscle. Acta Physiologica Scandinavica. 2005;**183**(1):43-58

[26] Wang L, Liu Q, Kitamoto T, Hou J, Qin J, Accili D. Identification of insulin-responsive transcription factors that regulate glucose production by hepatocytes. Diabetes. 2019;**68**(6):1156-1167

[27] den Boer MAM, Voshol PJ, Kuipers F, Romijn JA, Havekes LM. Hepatic glucose production is more sensitive to insulin-mediated inhibition than hepatic VLDL-triglyceride production. American Journal of Physiology. Endocrinology and Metabolism. 2006;**291**:1360-1364

[28] Hatting M, Tavares CDJ, Sharabi K, Rines AK, Puigserver P. Insulin regulation of gluconeogenesis. Annals of the New York Academy of Sciences. 2018;**1411**(1):21-35

[29] Charlton M, Nair KS. Protein metabolism in insulin-dependent diabetes mellitus. The Journal of Nutrition. 1998;**128**(2):323S-327S

[30] Fukagawa NK, Minaker KL, Rowe JW, Goodman MN, Matthews DE, Bier DM, et al. Insulin-mediated reduction of whole body protein breakdown. Dose-response effects on leucine metabolism in postabsorptive men. The Journal of Clinical Investigation. 1985;**76**(6):2306-2311

[31] Newgard CB, An J, Bain JR, Muehlbauer MJ, Stevens RD, Lien LF, et al. A branched-chain amino acid-related metabolic signature that differentiates obese and lean humans and contributes to insulin resistance. Cell Metabolism. 2009;**9**(4):311-326

[32] Tessari P, Nosadini R, Trevisan R, de Kreutzenberg S, Inchiostro S, Duner E, et al. Defective suppression by insulin of leucine-carbon appearance and oxidation in type 1, insulin-dependent diabetes mellitus. Evidence for insulin resistance involving glucose and amino acid metabolism. The Journal of Clinical Investigation. 1986;**77**(6):1797-1804

[33] Yoon MS. The emerging role of branched-chain amino acids in insulin resistance and metabolism. Nutrients. 2016;**8**(7):405

[34] Asghari G, Farhadnejad H, Teymoori F, Mirmiran P, Tohidi M, Azizi F. High dietary intake of branched-chain amino acids is associated with an increased risk of insulin resistance in adults. Journal of Diabetes. 2018;**10**(5):357-364

[35] Wang L, Liu Q, Kitamoto T, Hou J, Qin J, Accili D. Identification of insulin-responsive transcription factors that regulate glucose production by hepatocytes. Diabetes. 2019;**68**(6):1156-1167

[36] den Boer MAM, Voshol PJ, Kuipers F, Romijn JA, Havekes LM. Hepatic glucose production is more sensitive to insulin-mediated inhibition than hepatic VLDL-triglyceride production. American Journal of Physiology Endocrinology Metabolism. 2006;**291**:1360-1364

[37] Hatting M, Tavares CDJ, Sharabi K, Rines AK, Puigserver P. Insulin regulation of gluconeogenesis. In: Annals of the New York Academy of Sciences. Vol. 1411. Blackwell Publishing Inc.; 2018. pp. 21-35

[38] Kamagate A, Dong HH. FoxO1 integrates insulin signaling to VLDL production. Cell Cycle. 2008;**7**(20):3162-3170

[39] Kamagate A, Qu S, Perdomo G, Su D, Dae HK, Slusher S, et al. FoxO1 mediates insulin-dependent regulation of hepatic VLDL production in mice. The Journal of Clinical Investigation. 2008;**118**(6):2347-2364

[40] Kim DH, Zhang T, Lee S, Calabuig-Navarro V, Yamauchi J, Piccirillo A, et al. FoxO6 integrates insulin signaling with MTP for regulating VLDL production in the liver. Endocrinology. 2014;**155**(4):1255-1267

[41] Scherer T, Ohare J, Diggs-Andrews K, Schweiger M, Cheng B, Lindtner C, et al. Brain insulin controls adipose tissue lipolysis and lipogenesis. Cell Metabolism. 2011;**13**(2):183-194

[42] Iwen KA, Scherer T, Heni M, Sayk F, Wellnitz T, Machleidt F, et al. Intranasal insulin suppresses systemic but not subcutaneous lipolysis in healthy humans. The Journal of Clinical Endocrinology and Metabolism. 2014;**99**(2):E246-E251

[43] Saltiel AR. Insulin signaling in the control of glucose and lipid homeostasis. Handbook Experimental Pharmacology. 2016;**233**:51-71

[44] Charlton M, Nair KS. Protein metabolism in insulin-dependent diabetes mellitus. The Journal of Nutrition. 1998;**128**(2):323S-327S

[45] Fukagawa NK, Minaker KL, Rowe JW, Goodman MN, Matthews DE, Bier DM, et al. Insulin-mediated reduction of whole body protein breakdown. Dose-response effects on leucine metabolism in postabsorptive men. The Journal of Clinical Investigation. 1985;**76**(6):2306-2311

[46] Newgard CB, Ann J, Bain JR, Muehlbauer MJ, Stevens RD, Lien LF, et al. A branched-chain amino acid-related metabolic signature that differentiates obese and lean humans and contributes to insulin resistance. Cell Metabolism. 2009;**9**(4):311-326

[47] Tessari P, Nosadini R, Trevisan R, de Kreutzenberg S, Inchiostro S, Duner E, et al. Defective suppression by insulin of leucine-carbon appearance and

oxidation in type 1, insulin-dependent diabetes mellitus. Evidence for insulin resistance involving glucose and amino acid metabolism. The Journal of Clinical Investigation. 1986;77(6):1797-1804

[48] Yoon MS. The emerging role of branched-chain amino acids in insulin resistance and metabolism. Nutrients. 2016;**8**(7):405

[49] Asghari G, Farhadnejad H, Teymoori F, Mirmiran P, Tohidi M, Azizi F. High dietary intake of branched-chain amino acids is associated with an increased risk of insulin resistance in adults. Journal of Diabetes. 2018;**10**(5):357-364

[50] Petersen MC, Shulman GI. Mechanisms of insulin action and insulin resistance. Physiological Reviews. 2018;**98**(4):2133-2223

[51] Kaksonen M, Roux A. Mechanisms of clathrin-mediated endocytosis. Nature Reviews. Molecular Cell Biology. 2018;**19**(5):313-326

[52] Hall C, Yu H, Choi E. Insulin receptor endocytosis in the pathophysiology of insulin resistance. Experimental & Molecular Medicine. 2020;**52**(6):911-920

[53] Caro JF, Ittoop O, Pories WJ, Meelheim D, Flickinger EG, Thomas F, et al. Studies on the mechanism of insulin resistance in the liver from humans with noninsulin-dependent diabetes. Insulin action and binding in isolated hepatocytes, insulin receptor structure, and kinase activity. The Journal of Clinical Investigation. 1986;**78**(1):249-258

[54] Komatsu M, Takei M, Ishii H, Sato Y. Glucose-stimulated insulin secretion: A newer perspective. Journal of Diabetes Investigation. 2013;**4**(6):511-516

[55] Straub SG, Sharp GWG. Glucose-stimulated signaling pathways in biphasic insulin secretion. Diabetes/Metabolism Research and Reviews. 2002;**18**(6):451-463

[56] Wang J, Gu W, Chen C. Knocking down insulin receptor in pancreatic beta cell lines with lentiviral-small hairpin RNA reduces glucose-stimulated insulin secretion via decreasing the gene expression of insulin, GLUT2 and Pdx1. International Journal of Molecular Sciences. 2018;**19**(4)

[57] Solimena M, Schulte AM, Marselli L, Ehehalt F, Richter D, Kleeberg M, et al. Systems biology of the IMIDIA biobank from organ donors and pancreatectomised patients defines a novel transcriptomic signature of islets from individuals with type 2 diabetes. Diabetologia. 2018;**61**(3):641-657

[58] Lyon J, Manning Fox JE, Spigelman AF, Kim R, Smith N, O'Gorman D, et al. Research-focused isolation of human islets from donors with and without diabetes at the Alberta Diabetes Institute Islet Core. Endocrinology. 2016;**157**(2):560-569

[59] Deng S, Vatamaniuk M, Huang X, Doliba N, Lian MM, Frank A, et al. Structural and functional abnormalities in the islets isolated from type 2 diabetic subjects. Diabetes. 2004;**53**(3):624-632

[60] D'alessio D. The role of dysregulated glucagon secretion in type 2 diabetes. Diabetes, Obesity and Metabolism. 2011;**13**(Suppl. 1):126-132

[61] Philippe J, Knepel W, Waeber G. Insulin regulation of the glucagon gene is mediated by an insulin-responsive DNA element. Proceedings of the National Academy Science USA. 1991;**88**(16):7224-7227

[62] Eissele R, Göke R, Willemer S. Glucagon-like peptide-1 cells in the gastrointestinal tract and pancreas of rat, pig and man. European Journal of Clinical Investigation. 1992;**22**:283-291

[63] Holst JJ. The physiology of glucagon-like peptide 1. Physiological Reviews. 2007;**87**(4):1409-1439

[64] Tolhurst G, Heffron H, Lam YS, Parker HE, Habib AM, Diakogiannaki E, et al. Short-chain fatty acids stimulate glucagon-like peptide-1 secretion via the G-protein-coupled receptor FFAR2. Diabetes. 2012;**61**(2):364-371

[65] Lund A. On the role of the gut in diabetic hyperglucagonaemia. Danish Medical Journal. 2017;**64**(4):B5340

[66] Drucker DJ. Glucagon-like Peptide-1 and the islet β-cell: Augmentation of cell proliferation and inhibition of apoptosis. Endocrinology. 2003;**144**:5145-5148

[67] Drucker DJ, Nauck MA. The incretin system: Glucagon-like peptide-1 receptor agonists and dipeptidyl peptidase-4 inhibitors in type 2 diabetes. Lancet. 2006;**368**(9548):1696-1705

[68] Gloyn AL, Drucker DJ. Precision medicine in the management of type 2 diabetes. The Lancet Diabetes and Endocrinology. 2018;**6**(11):891-900

[69] Ottosson-Laakso E, Krus U, Storm P, Prasad RB, Oskolkov N, Ahlqvist E, et al. Glucose-induced changes in gene expression in human pancreatic islets: Causes or consequences of chronic hyperglycemia. Diabetes. 2017;**66**(12):3013-3028

[70] Segerstolpe Å, Palasantza A, Eliasson P, Andersson EM, Andréasson AC, Sun X, et al. Single-cell transcriptome profiling of human pancreatic islets in health and type 2 diabetes. CellMetabolism. 2016;**24**(4):593-607

[71] Marselli L, Piron A, Suleiman M, Colli ML, Yi X, Khamis A, et al. Persistent or transient human β cell dysfunction induced by metabolic stress: Specific signatures and shared gene expression with type 2 diabetes. Cell Reports. 2020;**33**(9):108466

[72] Marselli L, Thorne J, Dahiya S, Sgroi DC, Sharma A, Bonner-Weir S, et al. Gene expression profiles of Beta-cell enriched tissue obtained by laser capture microdissection from subjects with type 2 diabetes. PLoS One. 2010;**5**(7):e11499

[73] Weir GC, Bonner-Weir S. Five of stages of evolving β-cell dysfunction during progression to diabetes. Diabetes. 2004;**53**(Suppl. 3):S16-S21

[74] Khin PP, Lee JH, Jun HS. A brief review of the mechanisms of β-cell dedifferentiation in type 2 diabetes. Nutrients. 2021;**13**(5):1593

[75] Bensellam M, Jonas JC, Laybutt DR. Mechanisms of β;-cell dedifferentiation in diabetes: Recent findings and future research directions. The Journal of Endocrinology. 2018;**236**(2):R109-R143

[76] Sun T, Han X. Death versus dedifferentiation: The molecular bases of beta cell mass reduction in type 2 diabetes. Seminars in Cell & Developmental Biology. 2020;**103**:76-82

[77] Marrif HI, Al-Sunousi SI. Pancreatic β cell mass death. Frontiers in Pharmacology. 2016;**7**:83

[78] Rachdaoui N, Polo-Parada L, Ismail-Beigi F. Prolonged exposure to insulin inactivates Akt and Erk 1/2 and increases pancreatic islet and INS1E β-cell apoptosis. Journal of Endocrinology Society. 2018;**3**(1):69-90

[79] Bucris E, Beck A, Boura-Halfon S, Isaac R, Vinik Y, Rosenzweig T, et al. Prolonged insulin treatment sensitizes apoptosis pathways in pancreatic β cells. The Journal of Endocrinology. 2016;**230**(3):291-307

[80] Sampson SR, Bucris E, Horovitz-Fried M, Parnas A, Kahana S, Abitbol G, et al. Insulin increases H_2O_2-induced pancreatic beta cell death. Apoptosis. 2010;**15**(10):1165-1176

[81] Imai Y, Cousins RS, Liu S, Phelps BM, Promes JA. Connecting pancreatic islet lipid metabolism with insulin secretion and the development of type 2 diabetes. Annals of the New York Academy of Sciences. 2020;**1461**(1):53-72

[82] Lytrivi M, Castell AL, Poitout V, Cnop M. Recent insights into mechanisms of β-cell lipo- and glucolipotoxicity in type 2 diabetes. Journal of Molecular Biology. 2020;**432**(5):1514-1534

[83] Bensellam M, Jonas JC, Laybutt DR. Mechanisms of β-cell dedifferentiation in diabetes: Recent findings and future research directions. The Journal of Endocrinology. 2018;**236**(2):R109-R143

[84] Hunter CS, Stein RW. Evidence for loss in identity, De-differentiation, and trans-differentiation of islet β-cells in type 2 diabetes. Frontiers in Genetics. 2017;**8**:35

[85] Cinti F, Bouchi R, Kim-Muller JY, Ohmura Y, Sandoval PR, Masini M, et al. Evidence of β-cell dedifferentiation in human type 2 diabetes. The Journal of Clinical Endocrinology and Metabolism. 2016;**101**(3):1044-1054

[86] Brereton MF, Iberl M, Shimomura K, Zhang Q, Adriaenssens AE, Proks P, et al. Reversible changes in pancreatic islet structure and function produced by elevated blood glucose. Nature Communications. 2014;**5**:4639

[87] Wang Z, York NW, Nichols CG, Remedi MS. Pancreatic β cell dedifferentiation in diabetes and redifferentiation following insulin therapy. Cell Metabolism. 2014;**19**(5):872-882

[88] Neelankal John A, Ram R, Jiang FX. RNA-Seq analysis of islets to characterise the dedifferentiation in type 2 diabetes model mice db/db. Endocrine Pathology. 2018;**29**(3):207-221

[89] Talchai C, Xuan S, Lin H, Sussel L, Accili D. Pancreatic β cell dedifferentiation as a mechanism of diabetic β cell failure. Cell. 2012;**150**(6):1223-1234

[90] Spijker HS, Ravelli RBG, Mommaas-Kienhuis AM, van Apeldoorn AA, Engelse MA, Zaldumbide A, et al. Conversion of mature human β-cells into glucagon-producing α-cells. Diabetes. 2013;**62**(7):2471-2480

[91] Schuit F. Epigenetic programming of glucose-regulated insulin release. The Journal of Clinical Investigation. 2015;**125**(7):2565-2568

[92] Dhawan S, Tschen SI, Zeng C, Guo T, Hebrok M, Matveyenko A, et al. DNA methylation directs functional maturation of pancreatic β cells. The Journal of Clinical Investigation. 2015;**125**(7):2851-2860

[93] Spaeth JM, Walker EM, Stein R. Impact of Pdx1-associated chromatin modifiers on islet β-cells. Diabetes, Obesity & Metabolism. 2016;**18**(Suppl. 1):123-127

[94] Dayeh T, Ling C. Does epigenetic dysregulation of pancreatic islets

contribute to impaired insulin secretion and type 2 diabetes? Biochemistry and Cell Biology. 2015;**93**(5):511-521

[95] Campbell SA, Hoffman BG. Chromatin regulators in pancreas development and diabetes. Trends in Endocrinology and Metabolism. 2016;**27**(3):142-152

[96] Astro V, Adamo A. Epigenetic control of endocrine pancreas differentiation in vitro: Current knowledge and future perspectives. Frontiers in Cell and Development Biology. 2018;**6**:141

[97] Latreille M, Hausser J, Stützer I, Zhang Q, Hastoy B, Gargani S, et al. MicroRNA-7a regulates pancreatic β cell function. The Journal of Clinical Investigation. 2014;**124**(6):2722-2735

[98] Dumortier O, Fabris G, Pisani DF, Casamento V, Gautier N, Hinault C, et al. microRNA-375 regulates glucose metabolism-related signaling for insulin secretion. The Journal of Endocrinology. 2020;**244**(1):189-200

[99] Wan S, Zhang J, Chen X, Lang J, Li L, Chen F, et al. MicroRNA-17-92 regulates beta-cell restoration after streptozotocin treatment. Frontiers in Endocrinology. 2020;**11**:9

[100] Rodríguez-Comas J, Moreno-Asso A, Moreno-Vedia J, Martín M, Castaño C, Marzà-Florensa A, et al. Stress-induced microrna-708 impairs b-cell function and growth. Diabetes. 2017;**66**(12):3029-3040

[101] Fu X, Dong B, Tian Y, Lefebvre P, Meng Z, Wang X, et al. MicroRNA-26a regulates insulin sensitivity and metabolism of glucose and lipids. The Journal of Clinical Investigation. 2015;**125**(6):2497-2509

[102] Baroukh N, Ravier MA, Loder MK, Hill E, Bounacer A, Scharfmann R, et al. MicroRNA-124a regulates foxa2 expression and intracellular signaling in pancreatic β-cell lines. The Journal of Biological Chemistry. 2007;**282**(27):19575-19588

[103] Xu Y, Huang Y, Guo Y, Xiong Y, Zhu S, Xu L, et al. MicroRNA-690 regulates induced pluripotent stem cells (iPSCs) differentiation into insulin-producing cells by targeting Sox9. Stem Cell Research & Therapy. 2019;**10**(1):59

[104] Mahanjan A et al. Fine-mapping type 2 diabetes loci to single variant resolution using high-density imputation and islet specific epigenome maps. Nature Genetics. 2018;**50**:1505-1513

[105] Vujkovic M et al. Discovery of 318 new risk loci for type 2 diabetes and related vascular outcome among 1.4 million participants in a multi-ancestry meta-analysis. Nature Genetics. 2020;**52**:680-691

[106] De Jesus DF, Kulkarni RN. "Omics" and "epi-omics" underlying the β-cell adaptation to insulin resistance. Molecular Metabolism. 2019;**27**(Suppl):S42-S48

[107] Wortham M, Sander M. Mechanisms of β-cell functional adaptation to changes in workload. Diabetes, Obesity & Metabolism. 2016;**18**(Suppl 1):78-86

[108] Wortham M, Sander M. Transcriptional mechanisms of pancreatic β-cell maturation and functional adaptation. Trends in Endocrinology and Metabolism. 2021;**32**(7):474-487

[109] Kim H, Kulkarni RN. Epigenetics in β-cell adaptation and type 2 diabetes. Current Opinion in Pharmacology. 2020;**55**:125-131

[110] Kaviani M, Azarpira N, Karimi MH, Al-Abdullah I. The role of microRNAs in islet β-cell development. Cellular Biology International. 2016;**40**(12):1248-1255

[111] Eliasson L, Esguerra JLS. Role of non-coding RNAs in pancreatic beta-cell development and physiology. Acta Physiologica. 2014;**211**(2):273-284

[112] Kim KH, Hartig SM. Contributions of microRNAs to peripheral insulin sensitivity. Endocrinology. 2022;**163**(2):bqab250

[113] Jacovetti C, Abderrahmani A, Parnaud G, Jonas JC, Peyot ML, Cornu M, et al. MicroRNAs contribute to compensatory β cell expansion during pregnancy and obesity. The Journal of Clinical Investigation. 2012;**122**(10):3541-3551

[114] Cerf ME. Beta cell dynamics: Beta cell replenishment, beta cell compensation and diabetes. Endocrine;**44**(2):303-311

[115] Geach T. Diabetes: Preventing β-cell apoptosis and T2DM with microRNAs – A role for MIR-200? Nature Reviews Endocrinology. 2015;**11**(8):444

[116] Cinti F, Bouchi R, Kim-Muller JY, Ohmura Y, Sandoval PR, Masini M, et al. Evidence of β-cell dedifferentiation in human type 2 diabetes. The Journal of Clinical Endocrinology and Metabolism. 2016;**101**(3):1044-1054

[117] Khin PP, Lee JH, Jun HS. A brief review of the mechanisms of β-cell dedifferentiation in type 2 diabetes. Nutrients. 2021;**13**(5):1593

[118] Bensellam M, Jonas JC, Laybutt DR. Mechanisms of β-cell dedifferentiation in diabetes: Recent findings and future research directions. The Journal of Endocrinology. 2018;**236**(2):R109-R143

[119] Melkman-Zehavi T, Oren R, Kredo-Russo S, Shapira T, Mandelbaum AD, Rivkin N, et al. miRNAs control insulin content in pancreatic β-cells via downregulation of transcriptional repressors. EMBO Journal. 2011;**30**(5):835-845

[120] Bolmeson C, Esguerra JLS, Salehi A, Speidel D, Eliasson L, Cilio CM. Differences in islet-enriched miRNAs in healthy and glucose intolerant human subjects. Biochemical Biophysical Research Communication. 2011;**404**(1):16-22

[121] Nathan G, Kredo-Russo S, Geiger T, Lenz A, Kaspi H, Hornstein E, et al. MiR-375 promotes redifferentiation of adult human β cells expanded in vitro. PLoS One. 2015;**10**(4):e0122108-e0122108

[122] Xu G, Chen J, Jing G, Shalev A. Thioredoxin-interacting protein regulates insulin transcription through microRNA-204. Nature Medicine. 2013;**19**(9):1141-1146

[123] Sebastiani G, Po A, Miele E, Ventriglia G, Ceccarelli E, Bugliani M, et al. MicroRNA-124a is hyperexpressed in type 2 diabetic human pancreatic islets and negatively regulates insulin secretion. Acta Diabetologica. 2015;**52**(3):523-530

[124] Jo SH, Chen J, Xu G, Grayson TB, Thielen LA, Shalev A. miR-204 controls glucagon-like peptide 1 receptor expression and agonist function. Diabetes. 2018;**67**(2):256-264

[125] Brunton SA, Wysham CH. GLP-1 receptor agonists in the treatment of type 2 diabetes: Role and clinical experience to date. Postgraduate Medicine. 2020;**132**(suppl. 2):3-14

[126] Nachawi N, Rao PP, Makin V. The role of GLP-1 receptor agonists

in managing type 2 diabetes. Cleveland Clinic Journal of Medicine. 2022;**89**(8):457-464

[127] Li N, Jiang D, He Q, He F, Li Y, Deng C, et al. microRNA-181c-5p promotes the formation of insulin-producing cells from human induced pluripotent stem cells by targeting smad7 and TGIF2. Cell Death & Disease. 2020;**11**(6):1-12

[128] Bhushan R, Rani A, Gupta D, Akhtar A, Dubey PK. MicroRNA-7 regulates insulin signaling pathway by targeting IRS1, IRS2, and RAF1 genes in gestational diabetes mellitus. MicroRNA. 2022;**11**(1):57-72

[129] Rorsman P, Ashcroft FM. Pancreatic β-cell electrical activity and insulin secretion: Of mice and men. Physiological Reviews. 2018;**98**(1):117-214

[130] Ofori JK, Salunkhe VA, Bagge A, Vishnu N, Nagao M, Mulder H, et al. Elevated miR-130a/ miR130b/ miR-152 expression reduces intracellular ATP levels in the pancreatic beta cell. Scientific Reports. 2017;**7**(1):1-15

[131] Pullen TJ, da Silva XG, Kelsey G, Rutter GA. miR-29a and miR-29b contribute to pancreatic β-cell-specific silencing of Monocarboxylate transporter 1 (Mct1). Molecular and Cellular Biology. 2011;**31**(15):3182-3194

[132] Bagge A, Clausen TR, Larsen S, Ladefoged M, Rosenstierne MW, Larsen L, et al. MicroRNA-29a is up-regulated in beta-cells by glucose and decreases glucose-stimulated insulin secretion. Biochemical and Biophysical Research Communications. 2012;**426**(2):266-272

[133] Bagge A, Dahmcke CM, Dalgaard LT. Syntaxin-1a is a direct target of miR-29a in insulin-producing β-cells. Hormone and Metabolic Research. 2013;**45**(6):463-466

[134] Taskinen MR. Diabetic dyslipidemia. Atherosclerosis. Supplements. 2002;**3**(1):47-51

[135] Mahato RV, Gyawali P, Raut PP, Regmi P, Khelanand PS, Dipendra RP, et al. Association between glycaemic control and serum lipid profile in type 2 diabetic patients: Glycated haemoglobin as a dual biomarker. Biomedical Research. 2011;**22**(3):375-380

[136] Taskinen MR. Diabetic dyslipidaemia: From basic research to clinical practice. Diabetologia. 2003;**46**(6):733-749

[137] Tong X, Liu S, Stein R, Imai Y. Lipid droplets' role in the regulation of β-cell function and β-cell demise in type 2 diabetes. Endocrinology. 2022;**163**(3):bqac007

[138] Tarlton JMR, Patterson S, Graham A. MicroRNA sequences modulated by beta cell lipid metabolism: Implications for type 2 diabetes mellitus. Biology (Basel). 2021;**10**(6):534

[139] Lovis P, Roggli E, Laybutt DR, Gattesco S, Yang J-Y, Widmann C, et al. Alterations in microRNA expression contribute to fatty acid-induced pancreatic β-cell dysfunction. Diabetes. 2008;**57**:2728-2738

[140] Han Y-B, Wang M-N, Li Q, Guo L, Yang Y-M, Li P-J, et al. MicroRNA-34a contributes to the protective effects of glucagon-like peptide-1 against lipotoxicity in INS-1 cells. Chinese Medical Journal. 2012;**125**:4202-4208

[141] Lin X, Guan H, Huang Z, Liu J, Li H, Wei G, et al. Downregulation of Bcl-2 expression by miR-34a mediates

palmitate-induced Min6 cells apoptosis. Journal Diabetes Research. 2014;**2014**:258695

[142] Lu H, Hao L, Li S, Lin S, Lv L, Chen Y, et al. Elevated circulating stearic acid leads to a major lipotoxic effect on mouse pancreatic beta cells in hyperlipidaemia via a miR-34a-5p-mediated PERK/p53-dependent pathway. Diabetologia. 2016;**59**:1247-1257

[143] Kong X, Liu C-X, Wang G, Yang H, Yao X-M, Hua Q, et al. LncRNA LEGLTBC functions as a ceRNA to antagonize the effects of miR-34a on the downregulation of SIRT1 in glucolipotoxicity-induced INS-1 Beta cell oxidative stress and apoptosis. Oxidative Medicine and Cellular Longevity. 2019;**2019**:4010764

[144] Liu Y, Dong J, Ren B. MicroRNA-182-5p contributes to the protective effects of thrombospondin 1 against lipotoxicity in INS-1 cells. Experimental and Therapeutic Medicine. 2018;**16**:5272-5279

[145] Guo R, Yu Y, Zhang Y, Li Y, Chu X, Lu H, et al. Overexpression of miR-297b-5p protects against stearic acid-induced pancreatic β-cell apoptosis by targeting LATS2. American Journal of Physiology Metabolism. 2020;**318**:E430-E439

[146] Yu Y, Guo R, Zhang Y, Shi H, Sun H, Chu X, et al. miRNA-mRNA profile and regulatory network in stearic acid-treated β-cell dysfunction. The Journal of Endocrinology. 2020;**246**:13-27

[147] Li Y, Xu X, Liang Y, Liu S, Xiao H, Li F, et al. miR-375 enhances palmitate-induced lipoapoptosis in insulin-secreting NIT-1 cells by repressing myotrophin (V1) protein expression. International Journal of Clinical and Experimental Pathology. 2010;**3**:254-264

[148] Wang J, Lin Z, Yang Z, Liu X. lncRNA Eif4g2 improves palmitate-induced dysfunction of mouse β-cells via modulation of Nrf2 activation. Experimental Cell Research. 2020;**396**:112291

[149] Lovis P, Roggli E, Laybutt DR, Gattesco S, Yang JY, Widmann C, et al. Alterations in microRNA expression contribute to fatty acid-induced pancreatic beta-cell dysfunction. Diabetes. 2008;**57**(10):2728-2736

[150] Najafi-Shoushtari SH, Kristo F, Li Y, Shioda T, Cohen DE, Gerszten RE, et al. MicroRNA-33 and the SREBP host genes cooperate to control cholesterol homeostasis. Science. 2010;**328**(5985):1566-1569

[151] Han YB, Wang MN, Li Q, Guo L, Yang YM, Li PJ, et al. MicroRNA-34a contributes to the protective effects of glucagon-like peptide-1 against lipotoxicity in INS-1 cells. Chinese Medical Journal. 2012;**125**(23):4202-4208

[152] Lu H, Hao L, Li S, Lin S, Lv L, Chen Y, et al. Elevated circulating stearic acid leads to a major lipotoxic effect on mouse pancreatic beta cells in hyperlipidaemia via a miR-34a-5p-mediated PERK/p53-dependent pathway. Diabetologia. 2016;**59**(6):1247-1257

[153] Kong X, Liu C, Wang G, Yang H, Yao X, Hua Q, et al. LncRNA LEGLTBC functions as a ceRNA to antagonize the effects of miR-34a on the downregulation of SIRT1 in glucolipotoxicity-induced INS-1 beta cell oxidative stress and apoptosis. Oxidative Medicine and Cellular Longevity. 2019;**2019**:4010764

[154] Lin X, Guan H, Huang Z, Liu J, Li H, Wei G, et al. Downregulation of Bcl-2 expression by miR-34a mediates palmitate-induced Min6 cells

apoptosis. Journal Diabetes Research. 2014;**2014**:258695

[155] Bordone L, Motta MC, Picard F, Robinson A, Jhala US, Apfeld J, et al. Sirt1 regulates insulin secretion by repressing UCP2 in pancreatic beta cells. PLoS Biology. 2006;**4**(2):e31

[156] Solomon JM, Pasupuleti R, Lei X, McDonagh T, Curtis R, DiStefano PS, et al. Inhibition of SIRT1 catalytic activity increases p53 acetylation but does not alter cell survival following DNA damage. Molecular and Cellular Biology. 2006;**26**(1):28-38

[157] Lee J-H, Song M-Y, Song E-K, Kim E-K, et al. Overexpression of SIRT1 protects pancreatic β-cells against cytokine toxicity by suppressing the nuclear factor-κB signaling pathway. Diabetes. 2009;**58**(2):344-351

[158] Wu L, Zhou L, Lu Y, Zhang J, Jian F, Liu Y, et al. Activation of SIRT1 protects pancreatic β-cells against palmitate-induced dysfunction. Biochimica et Biophysica Acta. 2012;**1822**(11):1815-1825

[159] Lovis P, Roggli E, Laybutt DR, Gattesco S, Yang JY, Widmann C, et al. Alterations in microRNA expression contribute to fatty acid-induced pancreatic beta-cell dysfunction. Diabetes. 2008;**57**(10):2728-2736

[160] Liu Y, Dong J, Ren BO. MicroRNA-182-5p contributes to the protective effects of thrombospondin 1 against lipotoxicity in INS-1 cells. Experimental and Therapeutic Medicine. 2018;**16**(6):5272-5279

[161] Guo R, Yu Y, Zhang Y, Li Y, Chu X, Lu H, et al. Overexpression of miR-297b-5p protects against stearic acid-induced pancreatic β-cell apoptosis by targeting LATS2. American Journal of Physiology. Endocrinology and Metabolism. 2020;**318**(3):E430-E439

[162] Li Y, Xu X, Liang Y, Liu S, Xiao H, Li F, et al. miR-375 enhances palmitate-induced lipoapoptosis in insulin-secreting NIT-1 cells by repressing myotrophin (V1) protein expression. International Journal of Clinical and Experimental Pathology. 2010;**3**(3):254

[163] Paterson MR, Kriegel AJ. MiR-146a/b: A family with shared seeds and different roots. Physiological Genomics. 2017;**49**(4):243-252

[164] Fred RG, Bang-Berthelsen CH, Mandrup-Poulsen T, Grunnet LG, Welsh N. High glucose suppresses human islet insulin biosynthesis by inducing miR-133a leading to decreased polypyrimidine tract binding protein-expression. PLoS One. 2010;**5**(5):e10843

[165] Jiang W, Liu J, Dai Y, Zhou N, Ji C, Li X. MiR-146b attenuates high-fat diet-induced non-alcoholic steatohepatitis in mice. Journal of Gastroenterology and Hepatology. 2015;**30**(5):933-943

[166] Compagnoni C, Capelli R, Zelli V, Corrente A, Vecchiotti D, Flati I, et al. MiR-182-5p is upregulated in hepatic tissues from a diet-induced NAFLD/NASH/HCC C57BL/6J mouse model and modulates Cyld and Foxo1 expression. International Journal of Molecular Sciences. 2023;**24**:9239

[167] Kumar A, Ren Y, Sundaram K, Mu J, Sriwastva MK, Dryden GW, et al. miR-375 prevents high-fat diet-induced insulin resistance and obesity by targeting the aryl hydrocarbon receptor and bacterial tryptophanase (tnaA) gene. Theranostics. 2021;**11**(9):4061-4077